DRILLING AND COMPLETION IN PETROLEUM ENGINEERING

Multiphysics Modeling

Series Editors

Jochen Bundschuh
University of Southern Queensland (USQ),
Toowoomba, Australia
Royal Institute of Technology (KTH), Stockholm, Sweden

Mario César Suárez Arriaga
Department of Applied Mathematics and Earth Sciences,
Faculty of Physics and Mathematical Sciences, Michoacán University UMSNH,
Morelia, Michoacán, Mexico

ISSN: 1877-0274

Volume 3

Drilling and Completion in Petroleum Engineering
Theory and Numerical Applications

Editors

Xinpu Shen, Mao Bai & William Standifird

CRC Press
Taylor & Francis Group
Boca Raton London New York Leiden

CRC Press is an imprint of the
Taylor & Francis Group, an **informa** business

A BALKEMA BOOK

First issued in paperback 2017

CRC Press/Balkema is an imprint of the Taylor & Francis Group, an informa business

© 2012 Taylor & Francis Group, London, UK

Typeset by Vikatan Publishing Solutions (P) Ltd., Chennai, India

Published by: CRC Press/Balkema
                P.O. Box 447, 2300 AK Leiden, The Netherlands
                e-mail: Pub.NL@taylorandfrancis.com
                www.crcpress.com – www.taylorandfrancis.co.uk – www.balkema.nl

Library of Congress Cataloging-in-Publication Data

Drilling and completion in petroleum engineering: theory and numerical applications / editors, Xinpu Shen, Mao Bai & William Standifird.
    p. cm.
 (Multiphysics modeling, ISSN 1877-0274; v. 3)
 Includes bibliographical references and index.
 ISBN 978-0-415-66527-8 (hardback : alk. paper) -- ISBN 978-0-203-14472-5 (eBook)
1. Petroleum engineering—Mathematical models. I. Shen, Xinpu, 1963- II. Bai, Mao, 1952- III. Standifird, William.

TN870.53.D75 2012
622'.338201--dc23

                            2011027887

ISBN 13: 978-1-138-07386-9 (pbk)
ISBN 13: 978-0-415-66527-8 (hbk)

About the book series

Numerical modeling is the process of obtaining approximate solutions to problems of scientific and/or engineering interest. The book series addresses novel mathematical and numerical techniques with an interdisciplinary emphasis that cuts across all fields of science, engineering and technology. It focuses on breakthrough research in a richly varied range of applications in physical, chemical, biological, geoscientific, medical and other fields in response to the explosively growing interest in numerical modeling in general and its expansion to ever more sophisticated physics. The goal of this series is to bridge the knowledge gap among engineers, scientists, and software developers trained in a variety of disciplines and to improve knowledge transfer among these groups involved in research, development and/or education.

This book series offers a unique collection of worked problems in different fields of engineering and applied mathematics and science, with a welcome emphasis on coupling techniques. The book series satisfies the need for up-to-date information on numerical modeling. Faster computers and newly developed or improved numerical methods such as boundary element and meshless methods or genetic codes have made numerical modeling the most efficient state-of-the-art tool for integrating scientific and technological knowledge in the description of phenomena and processes in engineered and natural systems. In general, these challenging problems are fundamentally coupled processes that involve dynamically evolving fluid flow, mass transport, heat transfer, deformation of solids, and chemical and biological reactions.

This series provides an understanding of complicated coupled phenomena and processes, its forecasting, and approaches in problem solving for a diverse group of applications, including natural resources exploration and exploitation (e.g. water resources and geothermal and petroleum reservoirs), natural disaster risk reduction (earthquakes, volcanic eruptions, tsunamis), evaluation and mitigation of human induced phenomena (climate change), and optimization of engineering systems (e.g. construction design, manufacturing processes).

Jochen Bundschuh
Mario César Suárez Arriaga
(*Series Editors*)

Editorial board of the book series

Table of contents

Foreword

Location: Deepwater Gulf of Mexico
Objective: Drill past 26,400 ft to sub-salt petroleum reservoir

Drilling Foreman: "Are you sure this is going to work?"
Field Engineer: "Yes sir, my guys have run the simulation three times, all with the same results; we need to get this cross-linked polymer into the wellbore soon, or we will lose this hole section and have to bypass. The geomechanics engineers are confident that if we get this in place, we can restore full circulation and drill ahead."
Drilling Foreman: "OK, let's get on with it then. Driller, line up on the trip tank and call the mud man to mix the pill. I don't know about all this modeling, but we are out of options. If these guys are wrong, it won't be pretty, so let's make sure they stick around to see how this works out."

As a young field engineer, I didn't really know much about modeling, either. I had been trained as a measurements professional and had only recently begun to take some courses about how to apply those measurements to support operations. One of those courses included information about building simple geomechanical models of the Earth to calculate pore fluid and fracture pressures using real-time resistivity and sonic measurements. Unfortunately for the instructor, I remembered his name, and when I saw an opportunity to use a more sophisticated version of this geomechanical modeling in the field, I called him. Within a few hours, he was being whisked offshore to the wellsite by helicopter, something he had not done in almost 20 years.

In this case, the modeling worked well. We constructed and successfully applied a reasonably sophisticated mechanical simulation to hypothesis test several lost circulation mitigation solutions and to select the option with the highest probability of success. It does not always work out so well. Historical data reveals that, during well construction operations, up to 40% of the non-productive time experienced is related to preventable geomechanical issues; estimates of the value for this time are near $8B USD annually for the petroleum industry. This figure is likely dwarfed by the production losses resulting from sanding and other completion failures, but here the evidence is more anecdotal than factual for obvious commercial reasons. There are also environmental consequences and, perhaps most important, is that the safety of the men and women working in the field who depend on competent scientists and engineers to design and deploy barrier systems that will prevent unwanted flows from the wellbore.

At this point you might be thinking "Someone could make a company out of this geomechanical modeling business." Exactly, and that leads me to how I met the authors. In 2001, I left a large oilfield services provider for a start-up company (Knowledge Systems) that made geomechanical modeling software. We wanted to build a consulting business around the software tools, and set out to assemble a staff that could provide comprehensive geomechanical modeling solutions for our clientele. Unlike other firms, we chose to field deploy specialists who would operate real-time models built by our scientists. We used digital infrastructure to keep the field analysts connected to the scientists who would liaise with the customers onshore. This type of solution was radically different from the pre-drill static models in which the consultants were paid and gone by the time the customer determined whether or not any of the models matched reality. Our teams would literally be in the field, at the wellhead and accountable to the customers, for the quality of our predictions and any observed variances.

With so much at stake, I sought out the best in the industry; first, I found Dr. Mao Bai. In addition to a great education from the Pennsylvania State University, Dr. Bai brought our firm tremendous laboratory and modeling experience from the industry. He understands that making a great model is only the beginning; linking it to operations that can affect results is how modeling is put to work. Next, I hired Dr. Xinpu Shen, who is a world-class numerical modeler. Dr. Shen helped us to integrate basic analytical modeling with more advanced, dynamic simulations to solve increasingly complex customer challenges, such as casing collapse and salt subsidence. These gentlemen were major contributors to the success of the firm, so much so that it was acquired in 2008 by a major service provider. When these experts contacted me about writing a book, I thought who better than scientists who have worked in academia, the lab, and in the field. They know how the modeling works, and how to apply it to solve tangible challenges.

As you read this work, we want you to find more than just credible science and engineering. We want you to discover our experiences and the enthusiasm we have invested in creating innovative technical solutions to real world challenges. Most importantly, we hope you will take what we have learned, and then apply and expand on these techniques. Why? Well, our industry is counting on it. Demand is outstripping supply and reserves are becoming increasingly difficult to access and harvest. This means that we need strong technical solutions now more than ever to meet the world's energy demands.

William Standifird
Director – Solutions
Halliburton

About the editors

Xinpu Shen

Xinpu Shen, PhD, is a Principal Consultant in Halliburton Consulting, department of Petroleum Engineering, group of Geomechanics Practice. He received his PhD degree in Engineering Mechanics in 1994. He was an associate professor in Tsinghua University, Beijing, China, from 1994 to 1999. From 1997 to 2004, he worked as post-doctoral research associate in several European institutions include Politecnico di Milano, Italy, and the University of Sheffield, UK, etc. Since May 2001, he is a professor in Engineering Mechanics in Shenyang University of Technology, China. He worked as abroad consultant of geomechanics for Knowledge Systems Inc Houston since 2005 and until it is acquired by Halliburton in 2008. He has been coordinator to 3 projects supported by National Natural Science Foundation of China since 2005.

Mao Bai

Mao Bai is the Principal Consultant in Geomechanics and Geomechanics Solutions Team Leader in Halliburton Consulting and Project Management. Dr. Bai received the Master of Sciences degree from the University of Newcastle upon Tyne in mining engineering in U.K. in 1986, and Ph.D. from the Pennsylvania State University in mineral engineering in U.S.A. in 1991. Before joining Halliburton in 2008, Dr. Bai worked as a Senior Research Associate at the Rock Mechanics Institute in the University of Oklahoma between 1991 and 2000, as a Senior Engineer at TerraTek/Schlumberger between 2000 and 2007, and as a Senior Geomechanics Specialist at Geomechanics International/Baker Hughes between 2007 and 2008. Dr. Bai is specializing in technical advising in petroleum engineering related geomechanics. He is the author of the book "Coupled Processes in Subsurface Deformation, Flow and Transport" published by ASCE Press in 2000, and author/co-author of over 130 technical papers in geomechanics related subjects.

William Standifird

William Standifird currently serves as a Solutions Director for Halliburton. In this role he is charged with the invention, development and deployment of innovative technologies that support safe and efficient well construction for petroleum assets. William began his career with Schlumberger as a Drilling Services Engineer where he specialized in the application of petroleum geomechanics to deepwater drilling operations. He subsequently joined Knowledge Systems Inc. and rapidly built a global petroleum geomechanics practice which was acquired by Halliburton in 2008. William has over 20 peer reviewed publications, a Performed by Schlumberger Silver Medal and a Hart's Meritorious Engineering Award. He holds undergraduate degrees in electronics engineering, management science and earned a Master of Business Administration from the University of Houston System.

Acknowledgements

Warmest thanks are due to Professor Jochen Bundschuh and Professor Mario-César Suárez A, Editor and Co-editor of the series of *Multiphysics Modeling* of which this book is one volume, for very helpful academic communications from them.

Thanks are due to Ms Adrienne Silvan, Editor of Halliburton TPRB, for her careful revision on the draft of every chapters of this book.

Thanks are due to Professor Kaspar Willam from Houston University, and Professor Christine Ehlig-Economides from Texas A & M University, both members of National Academy of Engineering of US, for their kind helps during the preparation of this book.

Partial financial supports to the publication of this book from NNSF of China through contract (10872134) and from the Ministry of State Education of China through contract (208027), to Shenyang University of Technology, are gratefully acknowledged.

Xinpu Shen
May 2011

CHAPTER 1

Mathematical modeling of thermo-hydro-mechanical behavior for reservoir formation under elevated temperature

Xinpu Shen

1.1 INTRODUCTION

Oil and gas production and the related deformation of reservoir formations are typical examples of multi-physics phenomena existing in engineering. Liquid, gas, and solid skeleton are three kinds of continuum involved in the process, and each has its specific constitutive laws. In the process of oil and gas production, fluid flows within formation, and changes of pore pressure and other variables, such as temperature, will cause the deformation of the matrix. If enhancement is involved in the process of oil and gas production, mechanical damage will be the added mechanical variable, which must be modeled in the multi-physics description.

With a detailed analysis, liquid involved in a production process consists of oil and water; gas involved in production process consists of natural gas, carbon dioxide, and moisture vapor.

The thermo effect is important in the production enhancement process with hot steam injection. In this case, the desorption of both liquid oil and gas molecular from solid skeleton, along with the evaporation of oil and water could occur within the reservoir formation. This could result in a significant amount of thermal energy dissipation. Pore pressure variation will also be significant.

Because of the complexity of the multi-physics phenomena in the oil and gas production, it is difficult to perform a theoretical analysis without simplification.

The purpose of the simplifications adopted is to reduce the calculation burden without losing the accuracy of the description of major properties of the problems investigated.

With reference to the vast reported practices (Gregory and Turgay 1991; Kydland et al., 1988; Hassanizadeh 1986), in the process of drilling and completion, it is reasonable to simplify the multi-physics process into a process that consists of one type of liquid, one type of gas, and various kinds of solid materials. To simulate production enhancement, such as hydraulic fracturing and hot-steam injection, this kind of simplification is also believed to be reasonable.

This chapter aims to establish a mathematical model for fluid flow within a deformable porous continuum. Fluid will be treated as a mixture continuum of two phases that include gas and liquid. Physical processes, such as desorption, heat transfer and mass transfer, will be calculated in a coupled way with solid skeleton deformation. This model is a thermo-hydro-mechanical (THM) model; consequently, it can naturally account for the influence of pore pressure on the THM field variables.

There are two ways of deriving the finite element formulation with a set of governing differential equations: by variational principle or by weighted residual method. For the variational principle, it is necessary to find a scalar functional for the system studied, and the stationary condition of the functional will lead to its finite element formulation. Because the stiffness matrix obtained by the variational principle is usually symmetrical, the variational principle has obtained a wide application in deriving the finite element formulation. However, it is sometimes difficult to find a scalar functional for the system investigated; consequently, the variational principle cannot be used in these instances. Alternatively, the weighted residual method (Wand and Shoo 1992) is another good choice for deriving a finite element formulation. Based on the equivalent integral of the governing differential equations and its weak form, the weighted residual method uses a set of given functions to appropriate the field variables and to force the weighted sum of residuals of the governing equations with appropriate functions as solutions to be zero. Consequently, a set of finite element equations can be obtained. The weighted residual method does not require the scalar functional of the system. Therefore, for complex problems for which the scalar functional of the system is difficult to achieve, the weighted residual method was widely used to derive the finite element method (Zienkiewicz and Taylor 1993; Kohl *et al.*, 2004; Schaefer *et al.*, 2002; Zhou *et al.*, 1998). In this study, the Gale kin weighted residual method is adopted to derive the finite element formulation for the THM problems. The contents of this chapter include the following:

- Section 1.2 provides a general formulation and governing equations.
- Section 1.3 presents constitutive equations for the various physical processes of mass and heat transfer.
- Section 1.4 gives expressions for the variables used in this model, except the primary variables. Some of these expressions are based on experimental data and are expressed in terms of primary variables.
- Section 1.5 presents the resultant governing partial differential equations of the THM problems; boundary conditions are presented for the simplified governing differential equations.
- Section 1.6 presents the equivalent integral of the governing differential equation and its weak form.
- Section 1.7 presents the finite element formulation obtained with the Galerkin weighted residual method and its discretized form.
- Section 1.8 includes some remarks.

1.2 GENERAL CONSERVATION EQUATIONS OF HEAT AND MASS TRANSFER WITHIN A DEFORMABLE POROUS MEDIUM

The formulation of the THM model is derived on the basis of the Novier-Stokes equations (White 2003). The following assumptions are adopted in this model:

- The porous medium is homogeneous and isotropic.
- Local thermal equilibrium exists between different phases.
- Deformations are small, and strains are infinitesimal.
- The density of the solid phase is constant.
- The velocity of the solid phase is zero.
- The viscosity behavior of the solid phase is neglected.

1.2.1 *Macroscopic mass conservation equations*

For the solid phase, with reference to the Novier-Stokes equations (White 2003), the macroscopic mass conservation equation is written as:

$$\frac{D^s[(1-n)\rho^s]}{Dt} + (1-n)\rho^s \nabla \cdot \mathbf{v}^s + \dot{m}_{desorp} = 0 \tag{1.1}$$

where n is the porosity of the mixture and ρ^s is the density of solid skeleton; $\dot{m}_{desorp}$ is the change of solid density as a result of the desorption process, and it is assumed as a scalar. The symbol D/Dt is the Lagrange material derivative operator and is generally expressed as:

$$\frac{D}{Dt} = \frac{\partial}{\partial t} + \mathbf{v} \cdot \nabla, \quad \frac{D^\pi}{Dt} = \frac{\partial}{\partial t} + \mathbf{v}^\pi \cdot \nabla \tag{1.2}$$

where π indicates the quantity related to the π-phase with $\pi = s, l, g$, represents solid phase, liquid phase, and gas, respectively.

With the assumption of the zero velocity for the solid skeleton $\mathbf{v}^s = \mathbf{0}$, there is:

$$\frac{\partial[(1-n)\rho^s]}{\partial t} + \dot{m}_{desorp} = 0 \tag{1.3}$$

Consequently, there is:

$$\dot{m}_{desorp} = -\frac{\partial[(1-n)\rho^s]}{\partial t} \tag{1.4}$$

For liquid, the macroscopic mass conservation equation is:

$$\frac{D^l(nS_l\rho^l)}{Dt} + n\rho^l S_l \nabla \cdot \mathbf{v}^l - \dot{m}_{desorp} = 0 \tag{1.5}$$

where S_l is the volume fraction of liquid of the pore space (i.e., liquid saturation), ρ^l is density of the liquid. Eq. 1.5 can be alternatively written as:

$$\frac{\partial(nS_l\rho^l)}{\partial t} + \mathbf{v}^l \cdot \nabla(nS_l\rho^l) + n\rho^l S_l \nabla \cdot \mathbf{v}^l - \dot{m}_{desorp} = 0 \tag{1.6}$$

Under elevated temperature, two kinds of gaseous materials exist within the reservoir formation: natural gas and the gas evaporated from liquid. The mass conservation equation is shown in Eq. 1.7:

$$\frac{D^g\left(nS_g\rho^g\right)}{Dt} + nS_g\rho^g\nabla \cdot \mathbf{v}^g = 0 \tag{1.7}$$

where S_g is the volume fraction of the gaseous phase material in the pores, ρ^g is the density of the gas, and $\mathbf{v}^g$ is the velocity of the gas. Eq. 1.7 can be alternatively written as:

$$\frac{\partial\left(nS_g\rho^g\right)}{\partial t} + \mathbf{v}^g \cdot \nabla\left(nS_g\rho^g\right) + nS_g\rho^g\nabla \cdot \mathbf{v}^g = 0 \tag{1.8}$$

The following relationship is assumed for parameters S_l and S_g:

$$S_l + S_g = 1 \tag{1.9}$$

1.2.2   *Linear momentum conservation equations*

Because the volume fraction of the liquid and that of the gaseous material are rather small, it is reasonable to neglect them without losing the accuracy of the analysis of the THM problem. To further simplify the formulation, the volumetric force is also neglected in the study. No viscosity of solid phase material is considered here. Neglecting the influence of volume force on the deformation of the solid phase, as well as the linear momentum of liquid and of the gaseous materials, the conservation equation of the linear momentum is given as:

$$\nabla \cdot \boldsymbol{\sigma} = 0 \tag{1.10}$$

where $\boldsymbol{\sigma}$ is the second order nominal stress tensor.

1.2.3   *Energy (enthalpy) conservation equations*

A form of the energy conservation equation based on Navier-Stokes equation can be expressed as:

$$\rho\frac{DU}{Dt} = \frac{dW}{dt} - \nabla\cdot\mathbf{q} + \rho\frac{dR}{dt} \qquad (1.11)$$

where U is the internal energy of the system, which is proportional to its temperature; W is the work performed by external force; $\mathbf{q}$ is the heat flux. R is the phase-change (desorption/absorption)-resulted energy change of the solid phase, and ρ is the density of the material.

The following customary assumption as stated by White (2003) has been adopted:

$$dU \approx CdT, \quad \frac{\partial U}{\partial t} \approx \frac{\partial T}{\partial t}, \quad \rho,C,k,\mu \approx Const \qquad (1.12)$$

where C is the thermal capacity of the material, and T is the absolute temperature. Consequently, there is:

$$\rho\frac{DU}{Dt} = \rho\left(C\frac{\partial T}{\partial t} + C\mathbf{v}\cdot\nabla T\right) \qquad (1.13)$$

For the three-phase mixture, which models the reservoir formation as one continuum, the velocity of the solid phase is set to be zero, the velocity of the gaseous phase is $\mathbf{v}^g$, and the velocity of the liquid is $\mathbf{v}^l$. Applying Eq. 1.13 to each single phase and summing up the corresponding equations yields the following resultant energy conservation equation:

$$\rho\frac{DU}{Dt} = \left[(1-n)\rho^s C^s + n(S_l\rho^l C^l + S_g\rho^g C^g)\right]\frac{\partial T}{\partial t} + n(S_l\rho^l C^l\mathbf{v}^l + S_g\rho^g C^g\mathbf{v}^g)\cdot\nabla T \quad (1.14)$$

where ρ is the equivalent density of the three-phase mixture as a whole and is expressed as:

$$\rho = (1-n)\rho^s + n(S_l\rho^l + S_g\rho^g) \qquad (1.15)$$

The term related to phase change can be expressed as:

$$\rho\frac{dR}{dt} = \dot{m}_{desorp}\Delta H_{desorp} \qquad (1.16)$$

where ΔH_{desorp} is the energy released by the dehydration of unit density of the solid phase.

With the assumption that the heat resource is zero, the energy (enthalpy) conservation equation of the solid phase of a unit volume can be expressed as:

$$\left[(1-n)\rho^s C^s + n(S_l\rho^l C^l + S_g\rho^g C^g)\right]\frac{\partial T}{\partial t} + n(S_l\rho^l C^l\mathbf{v}^l + S_g\rho^g C^g\mathbf{v}^g)\cdot\nabla T$$
$$= \sigma_{ij}\frac{d\varepsilon_{ij}}{dt} - \nabla\cdot\mathbf{q} + \dot{m}_{desorp}\Delta H_{desorp} \qquad (1.17)$$

where σ_{ij} and ε_{ij} are the stress tensor and strain tensor, respectively. Neglecting the viscosity of the material yields:

$$\left[(1-n)\rho^s C^s + n(S_l\rho^l C^l + S_g\rho^g C^g)\right]\frac{\partial T}{\partial t} + n(S_l\rho^l C^l\mathbf{v}^l + S_g\rho^g C^g\mathbf{v}^g)\cdot\nabla T$$
$$+ \nabla\cdot\mathbf{q} + \frac{\partial\left[(1-n)\rho^s\right]}{\partial t}\Delta H_{desorp} = 0 \qquad (1.18)$$

1.3 CONSTITUTIVE LAWS

A constitutive equation is the mathematical relationship between a state variable (also known as an independent variable, which is directly measurable and corresponds uniquely to the thermodynamic state) and its conjugate forces (also known as a dependent variable, which is not directly measurable). If inelastic deformation is being assumed, an internal state variable, which is not directly measurable (such as plastic strain) will also appear. State variables adopted in the constitutive relationship are not the same as the primary variables used in the governing differential equations (GDE). Usually, the state variables involved in the mathematical formulation of a model consist of primary variables and secondary variables that are expressed in terms of primary variables. In this model, to simplify the formulation of the problem, internal variables related to mass and heat transfer, such as porosity, liquid saturation within the porous space, and desorption rate of the solid skeleton, are all calculated in an empirical manner.

The primary state variables adopted in this model include the following: displacement $\mathbf{u}$, temperature T, liquid pore pressure p, and gaseous pressure p^g. The other state variables and the internal variables related to mass and heat transfer will be given in the form of empirical expressions in terms of primary variables.

The fluxes of the fluid include the advective flow of gas and the advective flow of liquid. The constitutive equations for these processes are given in the following subsections.

1.3.1 *Constitutive equations for mass transfer*

1.3.1.1 *Advective flow of gas*
With the generalized Darcy's law to allow for relative permeability, the following is obtained for the advective mass flow of the gas:

$$\mathbf{v}^g = \frac{k^{rg}\mathbf{k}}{\mu^g}(-\nabla p^g + \rho^g \mathbf{g}) \tag{1.19}$$

where $\mathbf{v}^g$ is the mass flow velocity of gaseous phase (i.e., the quantity of mass transferred through a unit area in a unit time), p^g is gas pressure, ρ^g is the density of the gas phase, k^{rg} is the relative permeability for gas within reservoir formation, k is the intrinsic permeability of reservoir formation, $\mathbf{g}$ is gravity acceleration constant, and $|g| = 9.8$ m s^{-2}.

The *ideal gas assumption* is made for gas as a whole. With the Dalton's law for ideal gas, the following equations can be obtained:

For gaseous material in the reservoir, the following ideal gas laws are adopted:

$$p^g = \rho^g \frac{TR_0}{M_g} \tag{1.20}$$

where M_g is the molar mass of gas phase (kg · mol^{-1}). R_0 is the ideal gas constant, and its value is:

$$R = 8.314472 \quad \text{J·mol}^{-1}\text{·K}^{-1} \tag{1.21}$$

The equivalent molar mass of the gaseous phase can be obtained empirical expressions with reference to the contents of gas in practice.

1.3.1.2 *Advective mass flow of liquid*
With the generalized Darcy's law, which allows for relative permeability, the following is obtained for the advective flow of the liquid:

$$\mathbf{v}^l = \frac{k^{rl}\mathbf{k}}{\mu^l}(-\nabla p^l + \rho^l \mathbf{g}) \tag{1.22}$$

where $\mathbf{v}^l$ is the velocity of liquid, p^l is the liquid pore pressure, and ρ^l is the density of liquid. k^{rl} is the relative permeability of liquid within the reservoir formation.

1.3.2   *Constitutive equations for heat transfer*

1.3.2.1   *Conductive heat transfer within the domain* Ω
The heat flux can be obtained by Fourier's law:

$$\mathbf{q} = -\chi_{eff}\nabla T \tag{1.23}$$

where χ_{eff} is the effective thermal conductivity, which can be experimentally determined for the heat flow within the porous reservoir formation.

1.3.2.2   *Heat transferred in radiation at boundary* $\partial\Omega$

$$\mathbf{q} = e\sigma T^4 \tag{1.24}$$

where e is the emissivity $0 < e < 1$, σ is the Stefan-Bolzmann constant equal to 5.67×10^{-8} J/s $\cdot$ m$^2 \cdot$ K^4), and T is the absolute temperature.

1.3.3   *Constitutive equations for the mechanical response of the solid phase*

The strain decomposition can be made between the total strain and the strains corresponding to each phenomenon as follows:

$$\boldsymbol{\varepsilon} = \boldsymbol{\varepsilon}^e + \boldsymbol{\varepsilon}^p + \boldsymbol{\varepsilon}^{Th} + \boldsymbol{\varepsilon}^{lits} \tag{1.25}$$

where $\boldsymbol{\varepsilon}$, $\boldsymbol{\varepsilon}^e$, $\boldsymbol{\varepsilon}^p$, $\boldsymbol{\varepsilon}^{Th}$, $\boldsymbol{\varepsilon}^{lits}$ are the total strain tensor, elastic strain tensor, plastic strain tensor, thermal strain tensor, and the load induced thermal strain tensor (Khoury *et al.*, 1985), respectively. The load induced thermal strain was also called 'transient thermal creep.' The expression of these strain tensors in terms of the primary variables are given as follows:

$$\boldsymbol{\varepsilon} = \frac{1}{2}(\nabla\mathbf{u} + \nabla^T\mathbf{u}) \tag{1.26}$$

$$\boldsymbol{\varepsilon}^{Th} = \alpha(T - T_0)\mathbf{I} \tag{1.27}$$

$$\boldsymbol{\varepsilon}^p = \int_{t_0}^t \dot{\boldsymbol{\varepsilon}}^p dt \tag{1.28}$$

$$\boldsymbol{\varepsilon}^e = \boldsymbol{\varepsilon} - \boldsymbol{\varepsilon}^p \tag{1.29}$$

$$\boldsymbol{\varepsilon}^{lits} = \int_{t_0}^t \dot{\boldsymbol{\varepsilon}}^{lits} dt, \quad \dot{\boldsymbol{\varepsilon}}^{lits} = \frac{\beta}{f_c^0}\left[(1+\nu)\boldsymbol{\sigma}^- - \nu tr(\boldsymbol{\sigma}^-)\mathbf{I}\right]\dot{T}, \quad \dot{T} > 0 \tag{1.30}$$

where α is the thermal expansion coefficient, ν is Poisson's ratio, β is a constant, f_c^0 is uniaxial compressive strength, super script "-" indicates the part of compressive stress tensor, and $\mathbf{I}$ is the second order unit tensor.

The following elastic stress-strain relationship between the stress vector $\boldsymbol{\sigma}$ and the displacement vector $\mathbf{u}$ is assumed for the deformation of skeleton

$$\boldsymbol{\sigma}' = \mathbf{D} : (\boldsymbol{\varepsilon} - \boldsymbol{\varepsilon}^p - \boldsymbol{\varepsilon}^{Th} - \boldsymbol{\varepsilon}^{lits}) = \mathbf{D} : \left[\frac{1}{2}(\nabla\mathbf{u} + \nabla^T\mathbf{u}) - \alpha(T - T_0)\mathbf{I} - \boldsymbol{\varepsilon}^p - \boldsymbol{\varepsilon}^{lits}\right] \tag{1.31}$$

where **D** is the fourth order tangential stiffness tensor, which is stress-dependent for a nonlinear elastic problem, $\boldsymbol{\varepsilon}$ is the strain tensor, and α is the thermal expansion coefficient. The following relationship between the total nominal stress $\boldsymbol{\sigma}$ and the solid phase stress $\boldsymbol{\sigma}'$, which is also known as effective stress, has been adopted:

$$\boldsymbol{\sigma} = \mathbf{D} : \left[\frac{1}{2} \left(\nabla \mathbf{u} + \nabla^T \mathbf{u} \right) - \alpha \left(T - T_0 \right) \mathbf{I} - \boldsymbol{\varepsilon}^p - \boldsymbol{\varepsilon}^{lits} \right] - \left[p^g (1 - S_l) - S_l p^l \right] \mathbf{I} \tag{1.32}$$

1.4 SOME EMPIRICAL EXPRESSIONS

To calculate some of the secondary variables, empirical expressions in terms of primary variables are presented here with reference to the experimental data from references.

1.4.1 *The expression of total porosity n*

It is assumed that:

$$n = n_0 + A_n (T - T_0) \tag{1.33}$$

where A_n is a constant that is dependent on the type of formation. Here it is taken that:

$$A_n = 0.000195 \text{ K}{-}1 \tag{1.34}$$

The value $n_0 = 0.06$ is given for a general formation, and T_0 is the ambient temperature, which is set as:

$$T_0 = 293.15 \text{ K} \tag{1.35}$$

For temperatures below the complete evaporation point of the mixture liquid, it is important to distinguish between saturated and non-saturated (partially saturated) formations; for higher temperatures, however, there is no such distinction because the liquid phase does not exist at any temperature above the complete evaporation point.

1.4.2 *The expression of $\dot{m}_{desorp}$*

The following constant density of the solid phase is assumed in this model:

$$\rho^s = Const \tag{1.36}$$

Consequently, with Eq. 1.33 the following is obtained:

$$\frac{\partial \left[(1-n)\rho^s \right]}{\partial t} = -A_n \rho^s \frac{\partial T}{\partial t} \tag{1.37}$$

Therefore, it can be derived with Eq. 1.4 that:

$$\dot{m}_{desorp} = \frac{\partial m_{desorp}}{\partial t} = A_n \rho^s \frac{\partial T}{\partial t} \tag{1.38}$$

1.4.3 *Effective thermal conductivity of the three-phase medium*

The effective thermal conductivity (unit: [W · m^{-1} · K^{-1}]) of the reservoir formation as a three-phase mixture can be expressed as:

$$\chi_{eff} = 1.67\left[1 + 0.0005(T - 298.15)\right]\left[1 + \frac{4n\rho^l S_l}{(1-n)\rho^s}\right] \tag{1.39}$$

1.5 RESULTANT GOVERNING EQUATIONS

Substituting the constitutive relations derived in section 1.3 into the conservation equations listed in section 1.2, the following governing differential equations are obtained:

$$\frac{\partial(nS_g\rho^g)}{\partial t} + \left[\frac{k^{rg}k}{\mu^g}(-\nabla p^g + \rho^g\mathbf{g})\right] \cdot \nabla(nS_g\rho^g) + nS_g\rho^g\nabla \cdot \left[\frac{k^{rg}k}{\mu^g}(-\nabla p^g + \rho^g\mathbf{g})\right] = 0 \tag{1.40}$$

$$\frac{\partial(nS_l\rho^l)}{\partial t} + \left[\frac{k^{rl}k}{\mu^l}(-\nabla p^l + \rho^l\mathbf{g})\right] \cdot \nabla(nS_l\rho^l) + n\rho^l S_l\nabla \cdot \left[\frac{k^{rl}k}{\mu^l}(-\nabla p^l + \rho^l\mathbf{g})\right]$$
$$+ \frac{\partial\left[(1-n)\rho^s\right]}{\partial t} = 0 \tag{1.41}$$

$$\nabla \cdot \left\{\mathbf{D} : \left[\frac{1}{2}(\nabla\mathbf{u} + \nabla^T\mathbf{u}) - \alpha(T - T_0)\mathbf{I} - \boldsymbol{\varepsilon}^p - \boldsymbol{\varepsilon}^{lits}\right] - \left[p^g(1 - S_l) - S_l p^l\right]\mathbf{I}\right\} = \mathbf{0} \tag{1.42}$$

$$\left[(1-n)\rho^s C^s + n(S_l\rho^l C^l + S_g\rho^g C^g)\right]\frac{\partial T}{\partial t} + n\left\{S_l\rho^l C^l\left[\frac{k^{rl}k}{\mu^l}(\nabla p^l + \rho^l\mathbf{g})\right]\right.$$
$$\left. + S_g\rho^g C^g\left[\frac{k^{rg}k}{\mu^g}(-\nabla p^g + \rho^g\mathbf{g})\right]\right\} \cdot \nabla T - \nabla \cdot (\chi_{eff}\nabla T) + \frac{\partial\left[(1-n)\rho^s\right]}{\partial t}\Delta H_{desorp} = 0 \tag{1.43}$$

With the empirical expressions given in Section 1.4, the governing equations can be solved by using the finite element method with the given initial and boundary conditions.

The governing differential equations previously given have taken the gaseous phase pressure p^g, liquid pressure p^l, temperature T, and displacement vector $\mathbf{u}$ as primary variables.

At time $t = 0$, the initial conditions of the field variables are given over the domain Ω and its boundary $\partial\Omega$ as follows:

$$p^g\big|_{t=0} = p_0^g, \quad p^l\big|_{t=0} = p_0^l, \quad T\big|_{t=0} = T_0, \quad \mathbf{u}\big|_{t=0} = \mathbf{u}_0 \tag{1.44}$$

Boundary conditions at t > 0 on $\partial\Omega$ for those primary variables are given as follows:

1. For the first kind, also known as the Dirichlet's boundary conditions on Γ_1, there is:

$$p^g = \overline{p}^g, \quad p^l = \overline{p}^l, \quad T = \overline{T}, \quad \mathbf{u} = \overline{\mathbf{u}} \tag{1.45}$$

where the over bar indicates the variables given on the boundary.
2. For the second kind, also known as Neumann's conditions on $\Gamma2$, there are the following relationships for the three sets of conservation equations in Eq. 1.40, Eq. 1.41, and Eq. 1.43:

$$\boldsymbol{\sigma} \cdot \mathbf{n} = \overline{\mathbf{t}} \tag{1.46}$$

$$\mathbf{J}^g \cdot \mathbf{n} = \overline{q}^g \tag{1.47}$$

$$\mathbf{J}^l \cdot \mathbf{n} = \bar{q}^l \tag{1.48}$$

$$\{-\chi_{eff} \nabla T\} \cdot \mathbf{n} = \bar{q}^{Th} \tag{1.49}$$

where $\bar{q}^g$, $\bar{q}^l$, $\bar{q}^{Th}$ are the given values of fluxes of the gas, liquid and heat.

3. For the third kind of boundary conditions, also known as the Cauchy's mixed boundary conditions on $\Gamma 3$, the following relationship exists for the conservation equations in Eq. 1.43:

$$\left(-\chi_{eff} \nabla T\right) \cdot \mathbf{n} = \alpha_c \left(T - T_\infty\right) \tag{1.50}$$

where α_c is the convective heat exchange coefficient, and T_∞ is the environmental temperature.

There is also the following additive relationship for the boundary sections:

$$\partial \Omega = \Gamma_1 + \Gamma_2 + \Gamma_3 \tag{1.51}$$

1.6 EQUIVALENT INTEGRAL OF THE GOVERNING DIFFERENTIAL EQUATION AND ITS WEAK FORM

The weighted Galerkin residual method is adopted to establish a finite element formulation for the coupled THM problem, which consists of gas, liquid, and solid phase materials. Following the standard procedure of the finite element formulation, the formulation of the weighted integral statement for the governing differential equations and boundary conditions can be written as the following:

$$\int_\Omega w_g \left\{ \frac{\partial (nS_g \rho^g)}{\partial t} + \left[\frac{k^{rg}k}{\mu^g}(-\nabla p^g + \rho^g \mathbf{g}) \right] \cdot \nabla (nS_g \rho^g) \right.$$
$$\left. + nS_g \rho^g \nabla \cdot \left[\frac{k^{rg}k}{\mu^g}(-\nabla p^g + \rho^g \mathbf{g}) \right] \right\} d\Omega + \int_{\Gamma_2} \bar{w}_g \left[nS_g (\mathbf{J}^g \cdot \mathbf{n} - \bar{q}^g) \right] dA = 0 \tag{1.52}$$

$$\int_\Omega w_c \left\{ \frac{\partial (nS_l \rho^l)}{\partial t} + \left[\frac{k^{rl}k}{\mu^l}(-\nabla p^l + \rho^l \mathbf{g}) \right] \cdot \nabla (nS_l \rho^l) + n\rho^l S_l \nabla \cdot \left[\frac{k^{rl}k}{\mu^l}(-\nabla p^l + \rho^l \mathbf{g}) \right] \right.$$
$$\left. + \frac{\partial \left[(1-n)\rho^s \right]}{\partial t} \right\} d\Omega + \int_{\Gamma_2} \bar{w}_c \left[nS_l (\mathbf{J}^l \cdot \mathbf{n} - \bar{q}^l) \right] dA = 0 \tag{1.53}$$

$$\int_\Omega \mathbf{w}_u^T \cdot \left(\nabla \cdot \left\{ \mathbf{D} : \left[\frac{1}{2}(\nabla \mathbf{u} + \nabla^T \mathbf{u}) - \alpha (T - T_0)\mathbf{I} - \boldsymbol{\varepsilon}^p - \boldsymbol{\varepsilon}^{lits} \right] - \left[p^g(1 - S_l) - S_l p^l \right]\mathbf{I} \right\} \right) d\Omega$$
$$+ \int_{\Gamma_2} \bar{\mathbf{w}}_u^T \cdot \left(\boldsymbol{\sigma} \cdot \mathbf{n} - \bar{\mathbf{t}} \right) dA = 0 \tag{1.54}$$

$$\int_\Omega w_T \left(\left[(1-n)\rho^s C^s + n(S_l \rho^l C^l + S_g \rho^g C^g) \right] \frac{\partial T}{\partial t} + n \left\{ S_l \rho^l C^l \left[\frac{k^{rl}k}{\mu^l}(\nabla p^l + \rho^l \mathbf{g}) \right] \right. \right.$$
$$\left. \left. + S_g \rho^g C^g \left[\frac{k^{rg}k}{\mu^g}(-\nabla p^g + \rho^g \mathbf{g}) \right] \right\} \cdot \nabla T - \nabla \cdot (\chi_{eff} \nabla T) + \frac{\partial \left[(1-n)\rho^s \right]}{\partial t} \Delta H_{desorp} \right) d\Omega \tag{1.55}$$
$$+ \int_{\Gamma_2} \bar{w}_T \left[(-\chi_{eff} \nabla T) \cdot \mathbf{n} - \bar{q}^{Th} \right] dA + \int_{\Gamma_3} \bar{w}_T \left[(-\chi_{eff} \nabla T) \cdot \mathbf{n} - \alpha_c (T - T_\infty) \right] dA = 0$$

where w_g, w_l, w_T are the weighting coefficient functions for field variables p^g, p^l, T; $\mathbf{w}_u$ is the weighting function, a column vector, for the displacement vector. The number of components of $\mathbf{w}_u$ equals that of the displacement vector; $\bar{w}_g$, $\bar{w}_l$, $\bar{w}_T$ are the weighting coefficient functions for the residual terms that result from the boundary conditions.

The weak form of the integral statement of the governing equations can reduce the order of the governing differential equations from second order to first order; consequently, it can reduce the requirement on the continuity properties of the approximate functions. With the application of the Gauss-Green theorem, the weak form of the equivalent weighted integral statements of the governing differential equations is obtained as follows:

1. Mass conservation equation of gas:

$$\int_\Omega w_g \frac{\partial(nS_g\rho^g)}{\partial t}\,d\Omega + \int_{\partial\Omega}(w_g nS_g q^g)\,d\Omega + \int_{\Gamma_2}\bar{w}_g\left[nS_g(\mathbf{J}^g\cdot\mathbf{n}-\bar{q}^g)\right]dA$$
$$-\int_\Omega \nabla w_g\cdot(nS_g\rho^g)\left[\frac{k^{rg}k}{\mu^g}(-\nabla p^g+\rho^g\mathbf{g})\right]d\Omega = 0 \tag{1.56}$$

2. Mass conservation equation of liquid:

$$\int_\Omega w_c\left\{\frac{\partial(nS_l\rho^l)}{\partial t}+\frac{\partial\left[(1-n)\rho^s\right]}{\partial t}\right\}d\Omega - \int_\Omega \nabla w_c\cdot\left\{(n\rho^w S_w)\left[\frac{k^{rl}k}{\mu^l}(\nabla p^l+\rho^l\mathbf{g})\right]\right\}d\Omega$$
$$+\int_{\partial\Omega}w_c nS_l q^l\,dA + \int_{\Gamma_2}\bar{w}_c\left[nS_l(\mathbf{J}^l\cdot\mathbf{n}-\bar{q}^l)\right]dA = 0 \tag{1.57}$$

where

$$q^g = \rho^g\left[\frac{k^{rg}k}{\mu^g}(-\nabla p^g+\rho^g\mathbf{g})\right]\cdot\mathbf{n} \tag{1.58}$$

Vector $\mathbf{n}$ is the outward normal of the boundary surface, and

$$q^l = \rho^l\left[\frac{k^{rl}k}{\mu^l}(-\nabla p^l+\rho^l\mathbf{g})\right]\cdot\mathbf{n} \tag{1.59}$$

3. Momentum conservation equation:

$$-\int_\Omega \nabla\mathbf{w}_u^T:\left\{\mathbf{D}:\left[\frac{1}{2}(\nabla\mathbf{u}+\nabla^T\mathbf{u})-\alpha(T-T_0)\mathbf{I}-\boldsymbol{\varepsilon}^p-\boldsymbol{\varepsilon}^{lits}\right]-\left[p^g(1-S_l)-S_l p^l\right]\mathbf{I}\right\}d\Omega$$
$$+\int_{\partial\Omega}\mathbf{w}_u^T\,\boldsymbol{\sigma}\cdot\mathbf{n}\,dA+\int_{\Gamma_2}\bar{\mathbf{w}}_u^T\cdot(\boldsymbol{\sigma}\cdot\mathbf{n}-\bar{\mathbf{t}})\,dA = 0 \tag{1.60}$$

4. Energy conservation equation:

$$\int_\Omega w_T\left\{\left[(1-n)\rho^s C^s+n\,(S_l\rho^l C^l+S_g\rho^g C^g)\right]\frac{\partial T}{\partial t}+nS_g\rho^g C^g\left[\frac{k^{rg}k}{\mu^g}(-\nabla p^g+\rho^g\mathbf{g})\right]\cdot\nabla T\right.$$
$$\left.+nS_l\rho^l C^l\left[\frac{k^{rl}k}{\mu^l}(-\nabla p^l+\rho^l\mathbf{g})\right]\cdot\nabla T+\frac{\partial\left[(1-n)\rho^s\right]}{\partial t}\Delta H_{desorp}\right\}d\Omega$$
$$+\int_{\partial\Omega}w_T q^{Th}dA+\int_\Omega\nabla w_T\cdot(\chi_{eff}\nabla T)\,d\Omega-\int_{\Gamma_3}\bar{w}_T\left[(-\chi_{eff}\nabla T)\cdot\mathbf{n}-\alpha_c(T-T_\infty)\right]dA = 0 \tag{1.61}$$

where

$$q^{Th} = -\chi_{eff} \nabla T \cdot \mathbf{n} \tag{1.62}$$

The following choices have been made for the Galerkin weighting coefficients within the domain Ω and those on its boundary $\partial\Omega$:

$$w_i = -\bar{w}_i \tag{1.63}$$

Consequently, the set of conservation equations (from Eq. 1.56 to Eq. 1.61) can be rewritten in terms of w_i ($i = g, l, u, T$) as:

1. Mass conservation equation of gas:

$$\int_\Omega w_g \frac{\partial(nS_g\rho^g)}{\partial t} d\Omega + \int_{\Gamma_2} w_g nS_g \bar{q}^g dA - \int_\Omega \nabla w_g \cdot (nS_g\rho^g)\left[\frac{k^{rg}k}{\mu^g}(-\nabla p^g + \rho^g\mathbf{g})\right] d\Omega = 0 \tag{1.64}$$

2. Mass conservation equation of liquid:

$$\int_\Omega w_c \left\{\frac{\partial(nS_l\rho^l)}{\partial t} + \frac{\partial[(1-n)\rho^s]}{\partial t}\right\} d\Omega - \int_\Omega \nabla w_c \cdot \left\{(n\rho^l S_l)\left[\frac{k^{rl}k}{\mu^l}(-\nabla p^l + \rho^l\mathbf{g})\right]\right\} d\Omega \\ + \int_{\Gamma_2} w_c(nS_l\bar{q}^l)\, dA = 0 \tag{1.65}$$

3. Momentum conservation equation:

$$-\int_\Omega \nabla \mathbf{w}_u^T : \left\{\mathbf{D}:\left[\frac{1}{2}(\nabla\mathbf{u} + \nabla^T\mathbf{u}) - \alpha(T - T_0)\mathbf{I} - \boldsymbol{\varepsilon}^p - \boldsymbol{\varepsilon}^{lits}\right] - [p^g(1 - S_l) - S_l p^l]\mathbf{I}\right\} d\Omega \\ + \int_{\Gamma_2} \mathbf{w}_u^T \cdot \bar{\mathbf{t}}\, dA = 0 \tag{1.66}$$

4. Energy conservation equation:

$$\int_\Omega w_T \left\{\left[(1-n)\rho^s C^s + n\left(S_l\rho^l C^l + S_g\rho^g C^g\right)\right]\frac{\partial T}{\partial t} + nS_g\rho^g C^g\left[\frac{k^{rg}k}{\mu^g}(-\nabla p^g + \rho^g\mathbf{g})\right]\cdot\nabla T \right. \\ \left. + nS_l\rho^l C^l\left[\frac{k^{rl}k}{\mu^l}(-\nabla p^l + \rho^l\mathbf{g})\right]\cdot\nabla T + \frac{\partial[(1-n)\rho^s]}{\partial t}\Delta H_{desorp}\right\} d\Omega - \int_{\Gamma_2} w_T \bar{q}^{Th}\, dA \tag{1.67} \\ - \int_{\Gamma_3} w_T \alpha_c(T - T_\infty)dA + \int_\Omega \nabla w_T \cdot \left(\chi_{eff}\nabla T\right)d\Omega = 0$$

For brevity, after a series of mathematical operations, Eq. 1.56, Eq. 1.57, Eq. 1.60, and Eq. 1.61 can generally be written as follows:

1. Mass conservation equation of gas:

$$\int_\Omega w_g \left(c_{gg}\frac{\partial p^g}{\partial t} + c_{gT}\frac{\partial T}{\partial t}\right)d\Omega + \int_{\Gamma_2} w_g nS_g \bar{q}^g dA - \int_\Omega \nabla w_g \cdot (k_{gg}\nabla p^g + k_{gT}\nabla T + \mathbf{f}_g)\, d\Omega = 0 \tag{1.68}$$

2. Mass conservation equation of liquid:

$$\int_\Omega w_c \left(c_{cc} \frac{\partial p^l}{\partial t} + c_{cT} \frac{\partial T}{\partial t} \right) d\Omega + \int_\Omega \nabla w_c \cdot (k_{cc} \nabla p^l + k_{cT} \nabla T + \mathbf{f}_c) \, d\Omega + \int_{\Gamma_2} w_c (nS_l \bar{q}^l) \, dA = 0 \quad (1.69)$$

3. Momentum conservation equation:

$$\int_\Omega \nabla \mathbf{w}_u^T : \left\{ \mathbf{D} : \left[\frac{1}{2} (\nabla \mathbf{u} + \nabla^T \mathbf{u}) - \alpha (T - T_0)\mathbf{I} - \boldsymbol{\varepsilon}^p - \boldsymbol{\varepsilon}^{lits} \right] - \left[p^g (1 - S_l) - S_l p^l \right] \mathbf{I} \right\} d\Omega$$
$$- \int_{\Gamma_2} \mathbf{w}_u^T \cdot \bar{\mathbf{t}} dA = 0 \quad (1.70)$$

4. Energy conservation equation:

$$\int_\Omega w_T \left(c_{TT} \frac{\partial T}{\partial t} \right) d\Omega + \int_\Omega w_T \mathbf{k}_{TT} \cdot \nabla T d\Omega - \int_{\Gamma_2} w_T \bar{q}^{Th} dA - \int_{\Gamma_3} w_T \alpha_c T dA + \int_{\Gamma_3} w_T \alpha_c T_\infty dA = 0 \quad (1.71)$$

The coefficient terms displayed in Eq. 1.68 to Eq. 1.71 can be expressed as:

1. For mass conservation equation of gas:

$$c_{gg} = nS_g \frac{M_g}{TR} \quad (1.72)$$

$$c_{gT} = S_g \rho^g A_n - nS_g \frac{p^g M_g}{RT^2} \quad (1.73)$$

where

$$\frac{\partial m}{\partial T} = -\frac{44.68}{(22.34 + T')^2} \left(\frac{T_C + 10}{T_{C0} + 10} \right) \quad (1.74)$$

$$k_{gg} = -nS_g \rho^g \frac{k^{rg} k}{\mu^g} \quad (1.75)$$

$$\mathbf{f}_g = nS_g \rho^g \frac{k^{rg} k}{\mu^g} \rho^g \mathbf{g} \quad (1.76)$$

2. For mass conservation equation of liquid:

$$c_{cc} = -n\rho^l S_l \left(\frac{M_l}{\rho^l RTm} \right) \quad (1.77)$$

$$c_{cT} = (S_l \rho^l - \rho^s) A_n + nS_l \frac{\partial(\rho^l)}{\partial T} \quad (1.78)$$

$$k_{cc} = (n\rho^l S_l) \frac{k^{rl} k}{\mu^l} \quad (1.79)$$

$$\mathbf{f}_{cl} = -(n\rho^l S_l) \frac{k^{rl} k}{\mu^l} \rho^l \mathbf{g} \quad (1.80)$$

3. For energy conservation equation:

$$c_{TT} = \left[(1-n)\rho^s C^s + n(S_l \rho^l C^l + S_g \rho^g C^g)\right] - A_n \rho^s \Delta H_{desorp} \qquad (1.81)$$

$$\mathbf{k}_{TT} = nS_g \rho^g C^g \left[\frac{k^{rg}k}{\mu^g}(-\nabla p^g + \rho^g \mathbf{g})\right] + nS_l \rho^l C^l \left[\frac{k^{rl}k}{\mu^l}(-\nabla p^l + \rho^l \mathbf{g})\right] \qquad (1.82)$$

With Eq. 1.68 through Eq. 1.71 and a set of approximate functions for field variables, a finite element formulation for the THM problem is derived in the following section.

1.7　APPROXIMATE SOLUTION AND SPATIAL DISCRETIZATION

Using the Galerkin weighted residual method, the following approximations of the field variables are assumed:

$$p^g = \mathbf{N}_g \cdot \bar{\mathbf{p}}_g, \quad p^l = \mathbf{N}_l \cdot \bar{\mathbf{p}}_l, \quad T = \mathbf{N}_T \cdot \bar{\mathbf{T}}, \quad \mathbf{u} = \mathbf{N}_u \cdot \bar{\mathbf{u}} \qquad (1.83)$$

where

$$\mathbf{N}_g = \mathbf{N}_l = \mathbf{N}_T = \mathbf{N} \qquad (1.84)$$

where $\mathbf{N}$ indicates the shape function vector for the variables of pressure and temperature at each node; the symbol '·' indicates a scalar product operation. The expression of $\mathbf{N}$ is given in general as $\mathbf{N} = [N_i]$, $i = 1, n$ where n indicates the number of nodes; $\mathbf{N}_u$ indicates the second order tensor of the shape function for the displacement variables at each node, and is expressed as $\mathbf{N}_u = \left[N_{ij}^u\right]$ where subscript $i = 1, 3$ for a 3D problem, and $j = 1, n$. Corresponding to the assumption for the field variables in Eq. 1.83, the weighting functions in the Galerkin weighted residual formulation have also evolved into their vector form. The following forms for the weighting functions are assumed:

$$\mathbf{W}_g = \mathbf{W}_l = \mathbf{W}_T = \mathbf{N}, \quad \mathbf{W}_u = \mathbf{N}_u \qquad (1.85)$$

Substituting the above expressions given in Eq. 1.83 and Eq. 1.85 into the conservation equations (Eq. 1.68 to Eq. 1.71) yields the following equations:

1. For gas:

$$\int_\Omega \mathbf{N}^T \left(c_{gg}\mathbf{N} \cdot \dot{\bar{\mathbf{p}}}_g + c_{gT}\mathbf{N} \cdot \dot{\bar{\mathbf{T}}}\right) d\Omega + \int_{\Gamma_2} \mathbf{N}^T nS_g \bar{q}^g dA - \int_\Omega \nabla \mathbf{N}^T \cdot \left(k_{gg}\nabla \mathbf{N} \cdot \bar{\mathbf{p}}_g + \mathbf{f}_{g1}\right) d\Omega = 0 \quad (1.86)$$

2. For liquid:

$$\int_\Omega \mathbf{N}^T \left(c_{cc}\mathbf{N} \cdot \dot{\bar{\mathbf{p}}}_c + c_{cT}\mathbf{N} \cdot \dot{\bar{\mathbf{T}}}\right) d\Omega + \int_\Omega \nabla \mathbf{N}^T \cdot \left(k_{cc}\nabla \mathbf{N} \cdot \bar{\mathbf{p}}_l + \mathbf{f}_{c1}\right) d\Omega + \int_{\Gamma_2} \mathbf{N}^T \left(nS_l \bar{q}^l\right) dA = 0 \quad (1.87)$$

3. For momentum:

$$\int_\Omega \nabla \mathbf{N}_u^T : \left\{\mathbf{D} : \frac{1}{2}(\nabla \mathbf{N}_u \cdot \bar{\mathbf{u}} + \nabla^T (\mathbf{N}_u \cdot \bar{\mathbf{u}})) - \alpha(\mathbf{N} \cdot \bar{\mathbf{T}} - T_0)\mathbf{D} : \mathbf{I} \right. $$
$$\left. -\left[\mathbf{N} \cdot (1-S_l)\bar{\mathbf{p}}_g + S_l \mathbf{N} \cdot \bar{\mathbf{p}}_l\right]\mathbf{I}\right\} d\Omega - \int_\Omega \nabla \mathbf{N}_u^T : \left[\mathbf{D} : (\boldsymbol{\varepsilon}^p + \boldsymbol{\varepsilon}^{lts})\right] d\Omega - \int_{\Gamma_2} \mathbf{N}_u^T \cdot \bar{\mathbf{t}} dA = 0 \qquad (1.88)$$

For brevity, Eq. 1.88 can be alternatively written in the following form:

$$\int_{\Omega} \nabla \mathbf{N}_u^T : \left\{ \mathbf{D} : \mathbf{M} \cdot \bar{\mathbf{u}} + \alpha (\mathbf{N} \cdot \bar{\mathbf{T}} - T_0) \mathbf{D} : \mathbf{I} - \left[(1 - S_l) \mathbf{N} \cdot \bar{\mathbf{p}}_g + S_l \mathbf{N} \cdot \bar{\mathbf{p}}_l \right] \mathbf{I} \right\} d\Omega$$
$$- \int_{\Omega} \nabla \mathbf{N}_u^T : \left[\mathbf{D} : (\boldsymbol{\varepsilon}^p + \boldsymbol{\varepsilon}^{lits}) \right] d\Omega - \int_{\Gamma_2} \mathbf{N}_u^T \cdot \bar{\mathbf{t}} dA = 0 \qquad (1.89)$$

where $\mathbf{M}$ is a third order tensor, which can be expressed as:

$$M_{ijk} = \frac{1}{2} \left(\frac{\partial N_{ik}^u}{\partial x_j} + \frac{\partial N_{jk}^u}{\partial x_i} \right), \quad i, j = 1, 3; \quad k = 1, n \qquad (1.90)$$

4. For energy:

$$\int_{\Omega} \mathbf{N}^T (c_{TT} \mathbf{N} \cdot \dot{\bar{\mathbf{T}}}) d\Omega + \int_{\Omega} \nabla \mathbf{N}^T \cdot (k_{TT} \nabla \mathbf{N} \cdot \bar{\mathbf{T}}) d\Omega - \int_{\Gamma_2} \mathbf{N}^T \bar{q}^{Th} dA - \int_{\Gamma_3} \mathbf{N}^T \alpha_c (\mathbf{N} \cdot \bar{\mathbf{T}}) dA$$
$$+ \int_{\Gamma_3} \mathbf{N}^T \alpha_c T_\infty dA = 0 \qquad (1.91)$$

The set of integral statements of the weak-form conservation equations listed in Eq. 1.86 to Eq. 1.91 can be written in the following tensor form:

$$\mathbf{C} \cdot \dot{\mathbf{X}} + \mathbf{K} \cdot \mathbf{X} = \mathbf{F} \qquad (1.92)$$

where

$$\mathbf{C} = \begin{bmatrix} \mathbf{C}_{gg} & \mathbf{0} & \mathbf{C}_{gT} & \mathbf{0} \\ \mathbf{0} & \mathbf{C}_{cc} & \mathbf{C}_{cT} & \mathbf{0} \\ \mathbf{C}_{gT} & \mathbf{C}_{Tc} & \mathbf{C}_{TT} & \mathbf{0} \\ \mathbf{0} & \mathbf{0} & \mathbf{0} & \mathbf{0} \end{bmatrix} \qquad (1.93)$$

$$\mathbf{K} = \begin{bmatrix} \mathbf{K}_{gg} & \mathbf{0} & \mathbf{0} & \mathbf{0} \\ \mathbf{0} & \mathbf{K}_{cc} & \mathbf{0} & \mathbf{0} \\ \mathbf{0} & \mathbf{0} & \mathbf{K}_{TT} & \mathbf{0} \\ \mathbf{K}_{ug} & \mathbf{K}_{uc} & \mathbf{K}_{uT} & \mathbf{K}_{uu} \end{bmatrix} \qquad (1.94)$$

$$\mathbf{X} = \left\{ \bar{\mathbf{p}}_g \quad \bar{\mathbf{p}}_l \quad \bar{\mathbf{T}} \quad \bar{\mathbf{u}} \right\}^T \qquad (1.95)$$

where $\bar{\mathbf{p}}_g, \bar{\mathbf{p}}_l, \bar{\mathbf{T}}, \bar{\mathbf{u}}$ are vectors of nodal values of the primary variables p_g, p_l, T, u respectively. The right side term is:

$$\mathbf{F} = \left\{ \mathbf{F}_g \quad \mathbf{F}_l \quad \mathbf{F}_T \quad \mathbf{F}_u \right\}^T \qquad (1.96)$$

The coefficient tensors in Eq. 1.93 are given as:

$$\mathbf{C}_{gg} = \int_{\Omega} \mathbf{N}^T \otimes (c_{gg} \mathbf{N}) d\Omega \qquad (1.97)$$

$$\mathbf{C}_{gT} = \int_{\Omega} \mathbf{N}^T \otimes (c_{gT} \mathbf{N}) d\Omega \qquad (1.98)$$

$$\mathbf{C}_{cc} = \int_{\Omega} \mathbf{N}^T \otimes \left(c_{cc}\mathbf{N}\right) d\Omega \tag{1.99}$$

$$\mathbf{C}_{cT} = \int_{\Omega} \mathbf{N}^T \otimes \left(c_{cT}\mathbf{N}\right) d\Omega \tag{1.100}$$

$$\mathbf{C}_{Tc} = \int_{\Omega} \mathbf{N}^T \otimes \left(c_{Tc}\mathbf{N}\right) d\Omega \tag{1.101}$$

$$\mathbf{C}_{TT} = \int_{\Omega} \mathbf{N}^T \otimes \left(c_{TT}\mathbf{N}\right) d\Omega \tag{1.102}$$

$$\mathbf{K}_{gg} = -\int_{\Omega} \nabla \mathbf{N}^T \cdot \left(k_{gg}\nabla \mathbf{N}\right) d\Omega \tag{1.103}$$

$$\mathbf{K}_{cc} = \int_{\Omega} \nabla \mathbf{N}^T \cdot \left(k_{cc}\nabla \mathbf{N}\right) d\Omega \tag{1.104}$$

$$\mathbf{K}_{TT} = \int_{\Omega} \mathbf{N}^T \cdot \left(\mathbf{k}_{TT1} \cdot \nabla \mathbf{N}\right) d\Omega + \int_{\Omega} \nabla \mathbf{N}^T \cdot \left(k_{TT2}\nabla \mathbf{N}\right) d\Omega - \int_{\Gamma_3} \mathbf{N}^T \otimes \left(\alpha_c \mathbf{N}\right) dA \tag{1.105}$$

$$\mathbf{K}_{ug} = -\int_{\Omega} \nabla \mathbf{N}_u^T : \left(\mathbf{N}_g \otimes \mathbf{I}\right) d\Omega \tag{1.106}$$

$$\mathbf{K}_{uc} = \int_{\Omega} \nabla \mathbf{N}_u^T : \left(S_l \mathbf{N}_l \otimes \mathbf{I}\right) d\Omega \tag{1.107}$$

$$\mathbf{K}_{uu} = \int_{\Omega} \nabla \mathbf{N}_u^T : \left(\mathbf{D} : \mathbf{M}\right) d\Omega \tag{1.108}$$

$$\mathbf{K}_{uT} = \int_{\Omega} \left[\nabla \mathbf{N}_u^T : \left(\mathbf{D} : \mathbf{I}\right)\right] \otimes \left(\alpha \mathbf{N}\right) d\Omega \tag{1.109}$$

$$\mathbf{F}_g = \mathbf{F}_{g1} + \mathbf{F}_{g2} \tag{1.110}$$

where

$$\mathbf{F}_{g1} = \int_{\Omega} \nabla \mathbf{N}^T \cdot \mathbf{f}_{g1} d\Omega \tag{1.111}$$

$$\mathbf{F}_{g2} = -\int_{\Gamma_2} \mathbf{N}^T n S_g \bar{q}^g dA \tag{1.112}$$

$$\mathbf{F}_c = \mathbf{F}_{c1} + \mathbf{F}_{c2} \tag{1.113}$$

where

$$\mathbf{F}_{c1} = -\int_{\Omega} \nabla \mathbf{N}^T \cdot \mathbf{f}_{c1} d\Omega \tag{1.114}$$

$$\mathbf{F}_{c2} = -\int_{\Gamma_2} \mathbf{N}^T \left(n S_l \bar{q}^l\right) dA \tag{1.115}$$

$$\mathbf{F}_T = \mathbf{F}_{T1} + \mathbf{F}_{T2} \tag{1.116}$$

where

$$\mathbf{F}_{T1} = -\int_{\Omega} \nabla \mathbf{N}^T \cdot \mathbf{f}_{T1} d\Omega \tag{1.117}$$

$$\mathbf{F}_{T2} = \int_{\Gamma_2} \mathbf{N}^T \bar{q}^{Th} dA - \int_{\Gamma_3} \mathbf{N}^T \alpha_c T_\infty dA \tag{1.118}$$

$$\mathbf{F}_u = \int_\Omega \nabla \mathbf{N}_u^T : (\alpha T_0 \mathbf{D} : \mathbf{I}) \, d\Omega + \int_\Omega \nabla \mathbf{N}_u^T : \left[\mathbf{D} : (\boldsymbol{\varepsilon}^p + \boldsymbol{\varepsilon}^{lits}) \right] d\Omega + \int_{\Gamma_2} \mathbf{N}_u^T \cdot \bar{\mathbf{t}} dA \qquad (1.119)$$

The coefficient matrix in Eq. 1.93 is not symmetrical; consequently, mathematical programming techniques must be used to determine an approximate solution.

1.8 ENDING REMARKS

It is convenient to build a THM formulation by Galerkin weighted residual method for modeling the behavior of reservoir formation under elevated temperature. With the Onsager reciprocal principle, it is generally possible to have a finite element formulation for the THM problem with a symmetrical coefficient matrix. However, this will require many more parameters and will result in many additional computation tasks. The asymmetrical coefficient matrix has resulted from the simplification described in this chapter. However, there are currently several numerical techniques, such as mathematical programming, that can be effectively used for solving the set of equations that have an asymmetrical coefficient matrix. The computer implementation of the model presented is an important aspect for the application of the model, and it will be the task for the next stage of this work.

ACKNOWLEDGEMENTS

Partial financial support from China National Natural Science Foundation through contract 10872134, support from Liaoning Provincial Government through contract RC2008-125 and 2008RC38, support from the Ministry of State Education of China through contract 208027, and support from Shenyang Municipal Government's Project are gratefully acknowledged.

NOMENCLATURE

$\mathbf{D}$	=	Fourth order tangential stiffness tensor, Pa
k^{rl}	=	Relative permeability of liquid, Darcy, d
n	=	Porosity of the mixture
α	=	Thermal expansion coefficient,
v	=	Poisson's ratio
β	=	Coefficient constant
f_c^0	=	Uniaxial compressive strength, Pa
ρ^s	=	Density of solid skeleton, m/L^3, kg/m^3
$\dot{m}_{desorp}$	=	Change of solid density, m/L^3, kg/m^3
D/Dt	=	Lagrange material derivative operator
π	=	s, l, g, represents solid phase, liquid phase, and gas, respectively
χ_{eff}	=	Effective thermal conductivity, W $\cdot$ m^{-1} $\cdot$ K^{-1}
e	=	Stefan-Bolzmann constant, J/(s $\cdot$ m$^2 \cdot$ K^4)
$\mathbf{v}^s$	=	Velocity for the solid skeleton, L/t, m/s
S_l	=	Volume fraction of liquid of the pore space
ρ^l	=	Density of the liquid, m/L^3, kg/m^3
S_g	=	Volume fraction of the gaseous phase material in the pores
R_0	=	Ideal gas constant, J $\cdot$ mol$^{-1} \cdot$ K^{-1}
M_g	=	Molar mass of gas phase, kg $\cdot$ mol^{-1}
$\mathbf{g}$	=	Gravity acceleration constant,
p	=	Pore pressure, Pa

T	=	Temperature, K
u	=	Displacement vector
σ	=	Second order nominal stress tensor, F/L^2, Pa
ε	=	Second order strain tensor
ε^{lits}	=	Load induced thermal strain tensor
ε^{Th}	=	Thermal strain tensor
ε^{p}	=	Plastic strain tensor
ρ^{g}	=	Density of the gas, m/L^3, kg/m^3
v^{g}	=	Velocity of the gas, L/t, m/s
v^{l}	=	Velocity of the liquid, L/t, m/s
ρ	=	Equivalent density of the mixture as a whole, m/L^3, kg/m^3
ΔH_{desorp}	=	Energy released by the dehydration of unit density of the solid phase, *Joule*
U	=	Internal energy of the system, *Joule*
W	=	Work performed by external force, *Joule*
q	=	Heat flux,
C	=	Thermal capacity of the material,
R	=	Phase-change (desorption/absorption)-resulted energy change of the solid phase, *Joule*
w	=	Weighting coefficient functions for field variables
w	=	Weighting function for the vectors of field variables
n	=	Outward normal of the boundary surface
N	=	Shape function vector for the primary variables
THM	=	Thermo-hydro-mechanical
GDE	=	Governing differential equations

REFERENCES

Gregory, K. and Turgay, R.E.: State-of-the-art modeling for unconventional gas recovery. SPE 18947-PA. *SPE Journal of Formation Evaluation* 6:1 (1991), pp. 63–71.

Hassanizadeh, S.M.: Derivation of basic equations of mass transport in porous media, Part 2. Generalized Darcy's and Fick's laws. *Adv. Water Resources* 9:12 (1986), pp. 207–222.

Khoury, G.A., Gringer, B.N. and Sullivan, P.J.E.: Transient thermal strain of concrete: literature review, conditions within specimen and behaviour of individual constituents. *Mag. Concrete Res.* 37:132 (1985), pp. 131–144.

Kuhl, D., Bangert, F. and Meschke, G.: Coupled chemo-mechanical deterioration of cementitious materials. Part I: Modelling. *Inter. J. Solids Stru.* (2004) 41, pp. 15–40.

Kydland, T., Haugan, P.M., Bousquet, G. and Havig, S.O.: Application of unconventional techniques in constructing an integrated reservoir simulation model for Troll field. SPE 15623-PA. *SPE Journal of Reservoir Engineering* 3:3 (1988), pp. 967–976.

Schrefler, B.A., Khoury, G.A., Gawin, D. and Majorana, C.E.: Thermo-hydro-mechanical modelling of high performance reservoir formation at high temperatures. *Engng. Compu.* 19:7 (2002), pp. 787–819.

Wang, X. and Shao, M.: *Fundamental principle of finite element method and its applications.* 3rd Ed., Tsinghua University Press, Beijing, China 1992, (in Chinese).

White, F.M.: *Fluid mechanics.* 5th Ed., McGraw Hill, International Edition, London, 2003.

Zhou, Y., Rajapakse, R.K.N. and Graham, J.: A coupled thermo-poro-elastic model with thermo osmosis and thermal filtration. *Int. J. Solids Stru.* 24:23/24 (1998), pp. 3548–3572.

Zienkiewicz, O.C. and Taylor, R.E.: *The finite element method.* 4th Ed., Vol. II, McGraw-Hill, London, 1993.

CHAPTER 2

Damage model for rock-like materials and its application

Xinpu Shen & Jinlong Feng

2.1 INTRODUCTION

The inelastic failure of rock-like materials and structures is characterized by the initiation and evolution of cracks and the frictional sliding on the closed crack surfaces. Plastic damage models are the major measures used to address cracking-related failure analysis, and are widely used by various researchers, such as Lemaitre (1990); Chaboche (1992), Seweryn and Mroz (1998), and de Borst *et al.* (1999). As a typical rock-like material, the mechanical behavior of concrete at ambient temperature is similar to that of rock; its damage evolution behavior has been investigated widely. Several damage models for concrete-like materials are described in the existing literature.

One of the popular models used in practice is the Mazars-Pijaudier-Cabot damage model (Mazars and Pijaudier-Cabot 1989). However, because this damage model is expressed in a holonomic manner with respect to strain loading and no incremental form was presented for damage evolution law, the damage evolution is uncoupled with the evolution of plastic strain.

Another popular model is the damage model described by de Borst *et al.* (1999). In the de Borst *et al.* damage model, the damage evolution law is based on the total quantity of equivalent strain. In practice, a great deal of strength data is available in terms of equivalent stress and/or fracture energy, but very little data exists for strain-based criteria. Thus, the equivalent-strain-based damage model will require special tests. Consequently, to avoid additional experimental calibration of model parameters, a plasticity-based damage model with equivalent stress-based loading criterion is more practical for the simulation of the progressive failure of rock-like material than this one.

One of the plasticity-based damage models is the Barcelona model, which is reported by Lubliner *et al.* (1989), and adopted by Lee and Fenves (1998) and Nechnech *et al.* (2002). This model presents an holonomic relationship between damage and equivalent plastic strain. It also presents two damage variables that are designed for tensile damage and compressive damage.

An anisotropic damage model of fourth order tensor form is reported by Govindjee *et al.* (1995) for the simulation of plastic damage to concrete. This model was later extended by Meschke *et al.* (1998) to include Drucker-Prager types of plasticity and plastic damage. There are several advantages of this model. Because damage variable is implicitly included in the expression of elastic compliance tensor, there is no need to adopt the effective stress concept. Furthermore, the inelastic calculation can be performed in nominal stress space. In addition, there is no limit for any kind of anisotropy included in this model. This model, however, also includes several disadvantages. Meschke *et al.* (1998) showed that some weak points exist in the model with respect to its energy dissipation property. In addition, the proportion of plastic strain with respect to damage strain was designed to be governed by a constant parameter. Another disadvantage of the model is that it was only used for the analysis of mode-I fracture. Furthermore, it is actually difficult to connect the stress triaxiality dependent softening phenomena with any kind of fracture modes.

Anisotropic damage models of vector form and of second order tensor form were also investigated by several researchers (see Swoboda *et al.,* 1998). However, because of the difficulties in the simulation of the experimental phenomena of rock-like specimens, and its inconvenience regarding computational aspects, it is still not appropriate to adopt these second order damage tensors in the simulation of failure of reservoir formation.

Because of its simplicity and a reasonable capacity of problem representation, the isotropic damage model is the most popular damage model used in the simulation of the failure phenomena of rock-like materials and structures. Therefore, it is the choice of this chapter, despite of the facts that the damage of rock usually appears in the form of directional cracks and that the plasticity of concrete is often related to the confined frictional sliding between closed crack surfaces under ambient temperature. The anisotropic phenomena will be simulated by the scale damage model at the structural level, rather than at the material level.

The following sections introduce the damage model that was presented by Lubliner *et al.* (1996) and Lee and Fenves (1998), followed by an application of this damage model in the determination of the size of the damage process zone. This chapter also presents the experimental results of damage evolution with a set of four-point shearing beams. A comparison of the numerical results of damage evolution and the experimental data has been performed. Some remarks are provided at the end of the chapter.

2.2   THE BARCELONA MODEL: SCALAR DAMAGE WITH DIFFERENT BEHAVIORS FOR TENSION AND COMPRESSION

The Barcelona model was first proposed by Lubliner *et al.* (1996) and was improved upon by Lee and Fenves (1998). This model is a plasticity-based scalar continuum damage model. The mechanism for the damage evolution in this model includes two aspects: damage resulting from tensile cracking and damage resulting from compressive crushing. The evolution of plastic loading is controlled by two hardening parameters: the equivalent plastic strain $\bar{\varepsilon}_t^{pl}$ which is caused by tensile load, and the part $\bar{\varepsilon}_c^{pl}$, which is caused by compressive load.

2.2.1   *Uniaxial behavior of the Barcelona model*

As shown in Fig. 2.1(a), the material shows linear elastic behavior before the stress reaches the value σ_{t0}. Micro-crack/damage will begin as the stress values exceed the point of σ_{t0}. Strain-softening phenomenon appears as a result of damage evolution, and it will result in strain localization to the structure.

For the compressive behavior shown in Fig. 2.1 (b), the material also shows linear elastic behavior before the stress reaches the value σ_{c0}. Micro-crack/damage will begin as the stress values exceed the point of σ_{c0}. Strain-hardening appears and will last until stress level reaches σ_{cu}. As the stress level exceeds the point of σ_{cu}, the strain-softening phenomenon appears as a result of damage evolution, and it will result in strain localization to the structure.

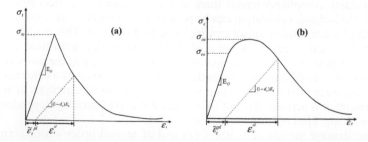

Figure 2.1.   Uniaxial behavior of the model: (a) under tension; (b) under compression.

Because of damage initiation and evolution, unloading stiffness is degraded from its original value of the intact material, as shown in Fig. 2.1. This stiffness degradation is expressed in terms of two damage variables: d_t and d_c. The value of the damage variables are 0 for the intact material and 1 for the completely broken state.

Assuming that E_0 is the Young's modulus of the initial intact material, the Hooke's law under uniaxial loading conditions will be:

$$\sigma_t = (1-d_t)E_0(\varepsilon_t - \bar{\varepsilon}_t^{pl}) \tag{2.1}$$

$$\sigma_c = (1-d_c)E_0(\varepsilon_c - \bar{\varepsilon}_c^{pl}) \tag{2.2}$$

Therefore, the effective stress for tension and compression can be written as:

$$\bar{\sigma}_t = \frac{\sigma_t}{1-d_t} = E_0(\varepsilon_t - \bar{\varepsilon}_t^{pl}) \tag{2.3}$$

$$\bar{\sigma}_c = \frac{\sigma_c}{1-d_c} = E_0(\varepsilon_c - \bar{\varepsilon}_c^{pl}) \tag{2.4}$$

Plastic yielding criteria will be described in the space of effective stress.

2.2.2   Unloading behavior

A description of the unloading behavior of the model is important for the application of the model to periodic loading conditions. The closure and opening of the existing micro-cracks will result in significant nonlinearity of the material behavior. It is experimentally proved that crack closure will result in some degree of stiffness recovery, which is also known as "unilateral effect."

The relationship of Young's modulus E for damaged material and that of the intact material E_0 is:

$$E = (1-d)E_0 \tag{2.5}$$

Lemaitre's "strain equivalent assumption" is adopted in this chapter. In Eq. 2.5, d is the synthetic damage variable, which is a function of stress state s, tensile damage d_t, and compressive damage d_c, and can be expressed as:

$$(1-d) = (1-s_t d_c)(1-s_c d_t) \tag{2.6}$$

where s_t and s_c are functions of the stress state and are calculated in the following manner:

$$s_t = 1 - w_t \gamma^*(\sigma_{11}); \qquad 0 \le w_t \le 1 \tag{2.7}$$

$$s_c = 1 - w_c(1 - \gamma^*(\sigma_{11})); \qquad 0 \le w_c \le 1 \tag{2.8}$$

In Eq. 2.7, $\gamma^*(\sigma_{11}) = H(\sigma_{11}) = \begin{cases} 1 & \text{as} \quad \sigma_{11} > 0 \\ 0 & \text{as} \quad \sigma_{11} < 0 \end{cases}$.

The weight parameters, w_t and w_c, are the material properties that control the amount of stiffness recovery for the unloading process, as shown in Fig. 2.2.

Hooke's law at the triaxial stress state is expressed in the following tensor form:

$$\boldsymbol{\sigma} = (1-d)\mathbf{D}_0^{el} : (\boldsymbol{\varepsilon} - \boldsymbol{\varepsilon}^{pl}) \tag{2.9}$$

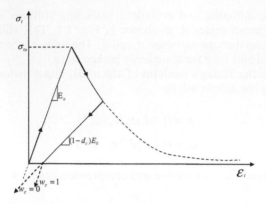

Figure 2.2.   The unloading behavior of the Barcelona model.

where $\mathbf{D}_0^{el}$ is the matrix of stiffness of the intact materials. At a triaxial stress state, the Heaviside function in the expression of the synthetic damage variable d can be written as the following form:

$$\gamma(\hat{\boldsymbol{\sigma}}) = \frac{\sum\limits_{i=1}^{3} <\sigma_i>}{\sum\limits_{i=1}^{3} |\hat{\sigma}_i|} \; ; \quad 0 \le \gamma(\hat{\boldsymbol{\sigma}}) \le 1 \tag{2.10}$$

where $\hat{\sigma}_i \; (i = 1, 2, 3)$ is the principal stress components, and the symbol $<\cdot>$ indicates that $<\chi> = \dfrac{1}{2}(|\chi| + \chi)$.

2.2.3 *Plastic flow*

The non-associated plastic flow rule is adopted in the model. The plastic potential G is in the form of Drucker-Prager type and is expressed as:

$$G = \sqrt{(\varepsilon \sigma_{t0} tg\psi)^2 + \bar{q}^2} - \bar{p} tg\psi \tag{2.11}$$

where $\psi(\theta, f_i)$ is the dilatancy angle, and $\sigma_{t0}(\theta, f_i) = \sigma_t |_{\bar{\varepsilon}_t^{pl}=0}$ is the threshold value of the tensile stress at which damage initiates. Parameter $\varepsilon(\theta, f_i)$ is a model parameter that defines the eccentricity of the loading surface in the effective stress space.

2.2.4 *Yielding criterion*

The yielding criterion of the model is given in the effective stress space, and its evolution is determined by two variables $\bar{\varepsilon}_t^{pl}$ and $\bar{\varepsilon}_c^{pl}$. Its expression is provided in Eq. 2.12:

$$F = \frac{1}{1-\alpha}(\bar{q} - 3\alpha\bar{p} + \beta(\bar{\varepsilon}^{pl}) <\hat{\bar{\sigma}}_{max}> -\gamma <-\hat{\bar{\sigma}}_{max}>) - \bar{\sigma}_c(\bar{\varepsilon}_c^{pl}) = 0 \tag{2.12}$$

where

$$\alpha = \frac{(\sigma_{b0}/\sigma_{c0}) - 1}{2(\sigma_{b0}/\sigma_{c0}) - 1}, 0 \le \alpha \le 0.5; \beta = \frac{\bar{\sigma}_c(\bar{\varepsilon}_c^{pl})}{\bar{\sigma}_t(\bar{\varepsilon}_t^{pl})}(1-\alpha) - (1+\alpha); \gamma = \frac{3(1-k_c)}{2k_c - 1} \tag{2.13}$$

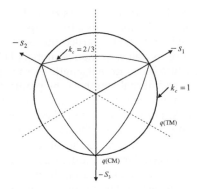

Figure 2.3.   The yielding surface for various values of k_c at π plane.

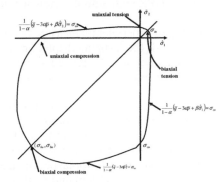

Figure 2.4.   The yielding surface for plane stress state.

where $\hat{\bar{\sigma}}_{max}$ is the value of maximum principal stress component; σ_{b0}/σ_{c0} is the ratio between the limit of bi-axial compression and that of uniaxial compression. Parameter k_c is the ratio between the second invariant of the stress tensor at the tensile meridian $q(TM)$ and the second invariant at the compressive meridian $q(CM)$ under arbitrary pressure which makes $\hat{\bar{\sigma}}_{max} < 0$. It is required that $0.5 < k_c \leq 1.0$, with a default value as 2/3. In Eq. 2.12, $\bar{\sigma}_t(\bar{\varepsilon}_t^{pl})$ is the effective tensile strength and $\bar{\sigma}_c(\bar{\varepsilon}_c^{pl})$ is the effective compressive strength. Fig. 2.3 shows the yielding surface for various values of k_c at π-plane. Fig. 2.4 shows the yielding surface for the plane-stress state.

2.3   CALIBRATION FOR THE SIZE OF DAMAGE PROCESS ZONE

Because of the heterogeneous material property of concrete-like materials, non-local inelastic models are becoming increasingly popular for addressing fractures and damage of concrete structures (Bazant and Pijaudier-Cabot 1988; Aifantis 1992 and 2003; Saanouni *et al.,* 1988). The internal length is an important parameter of a gradient-enhanced, non-local damage model and of an area-averaged, non-local model. In a gradient-enhanced non-local model, such as that described in Aifantis (1992), the internal length is the parameter that controls influence of its gradient enhancement. In an area-averaged, non-local model, the internal length is the parameter that defines the scope of the averaging calculation.

However, the definition of the internal length for a non-local model for concrete-like material has never been explicitly given. Consequently, the calibration of the internal length has not been effectively investigated. The damage process zone (DPZ) is the region in which the material degradation occurs before the macro-fracture appears within concrete-like quasi-brittle materials. Bazant and Cedolin (1991) regarded the length of the DPZ as the internal length, but did not explicitly give its value. Some researchers believe that the width of the crack that occurs within a structure is its internal length. For the gradient-enhanced damage model, Shen *et al.* (2005) used the internal length as the parameter that indicates the influence of the damage-gradient enhancement term on the non-local behavior and proposed its value in a curve-fitting manner. Although many researchers have used the concept of internal length, few have made explicit statements about the determination of its value.

With reference to the principle of 'non-local energy dissipation,' it is believed here that the internal length of a gradient-enhanced damage model should be the length of the DPZ. The damage process at points within the same DPZ can influence one another, and it will not be influenced by the energy value outside this damage process zone. Consequently, the length of a DPZ can represent the influence scope of a damage process; it should be regarded as the internal length of a non-local damage model. The goal of this chapter is to experimentally measure the length of the maximum DPZ with a double-notched, four-point shear concrete beam.

The white-light speckle method is an experimental measure that is widely used to measure surface deformation. With this method, the in-plane displacement field can be recorded at every time point within a given time interval. The related strain field can be calculated on the basis of the difference of the displacement field by comparing two displacement fields at different time points.

The white-light speckle method can be used to record the displacement field and to derive the related strain field that occurs on the surface of a specimen; however, it cannot measure the damage process zone, which is 'hidden'/invisible and located in front of an observable macro-crack. The damage process can only be calculated numerically with a set of given elastoplastic damage constitutive relations. The elastoplastic damage constitutive model used in this chapter was proposed by Lubliner *et al.* (1989) and further developed by Fenves and Lee (1998).

The following sections first introduce experiments performed with the white-light speckle method and four-point shear. The numerical results obtained with the finite-element analysis will be presented afterwards, followed by comparisons of the experimental and numerical results.

2.3.1 *Experiments performed with the white-light speckle method and four-point shear beam*

2.3.1.1 *Testing device*
Figure 2.5 shows the testing system used for these experiments. A concentrated loading force, P, is applied on the loading beam, which is made of steel. Force, P, is applied on the concrete specimen through two rollers, which are set in a way to redistribute it into force P1 and force P2 at the roller positions, with P2 equal to P1/15. P1 and the reaction force at the inside supporting roller will form a narrow shearing region within, where material points will be at a shear-stress state. The geometrical parameters of the specimen include the following: height of 150 mm, length of 400 mm. The width of the notch is 5 mm with a depth of 25 mm. The horizontal distance between P1 and the contact point of the right supporting roller is 25 mm, and the distance from the contact point between the left supporting roller and the left edge of the specimen is 12.5 mm.

2.3.1.2 *Experimental results*
Figure 2.6 shows the displacement-force diagram of the full-loading process. The total process was recorded by 13,500 digital photos; the camera speed is 15 pictures per second. Fig. 2.7 shows four resulting pictures of maximum shear strain, γ_{max}.

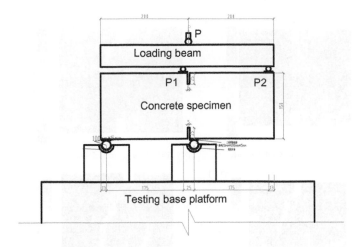

Figure 2.5.   Geometry of the loading system.

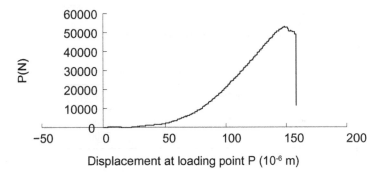

Figure 2.6.   Loading force vs. displacement at loading point P.

Strains shown in Fig. 2.7 (a) through (d) were calculated in terms of the displacement field recorded by the white-light speckle method. Fig. 2.7 indicates the range of the localization zone of the shear strain (i.e., shear band) corresponding to various loading time steps.

Fig. 2.7 (a) was taken at a stage that had no obvious damage and strain localization. Fig. 2.7 (b) was taken at the stage in which the localization band had just been formed. Fig. 2.7 (c) was taken at the moment that the macro-fracture had just formed. Fig. 2.7 (d) illustrates the moment when the secondary DPZ was formed. As previously discussed, although there are obvious localization bands formed within the specimen, it is not possible to determine where the DPZ begins and ends without a reference to a specific plastic damage constitutive model. This problem will be solved in the following section by using finite-element analysis and the specific damage model known as the Barcelona model.

2.3.2   *Numerical results obtained with finite-element analysis*

In the numerical study, two models were used with reference to the experimental devices. The first model is the four-point, double-notched shear beam; the second model is the single-notched, four-point shear beam. The parameters for the double-notched beam are the same as the specimen used in the experiment described in the previous section. The numerical results of strain obtained with the models used here are compared with the results obtained

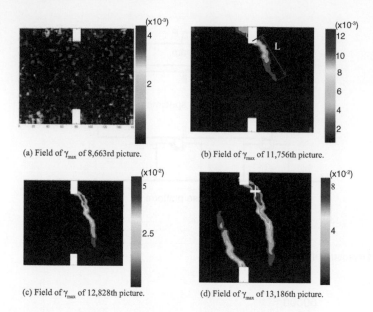

(a) Field of γ_{max} of 8,663rd picture.   (b) Field of γ_{max} of 11,756th picture.

(c) Field of γ_{max} of 12,828th picture.   (d) Field of γ_{max} of 13,186th picture.

Figure 2.7.   Evolution of field of γ_{max} and damage process zone.

Table 2.1.   Values of elastic parameters and density.

	E/Pa	v	ρ/(kg/m^3)
Concrete	2.72e+10	0.18	2400
Steel	2.0e+11	0.3	7800

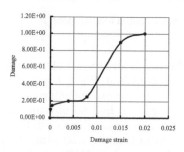

Figure 2.8.   Data that defines the evolution of tensile damage d_t.

by specimen testing. The model of single-notched, four-point beam is used to determine the variation of the geometrical character of a DPZ with a different specimen.

Table 2.1 lists the elasticity and density values. The values of the plastic parameters are given in the following: dilatancy angle $\psi = 38.0°$; eccentricity of plastic potential $\varepsilon = 0.1$, $\sigma_{b0}/\sigma_{c0} = 1.16$ $K_c = 0.7$.

The parameter values that define the evolution of tensile damage d_t are shown in Fig. 2.8, and the parameter values that define compressive damage d_c are shown in Fig. 2.9.

Fig. 2.10 illustrates the data that defines the plastic strain hardening–softening under compression. Fig. 2.11 illustrates the data that defines the plastic softening under tensional loading.

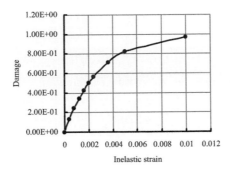

Figure 2.9. Data that defines the evolution of compressive damage d_c.

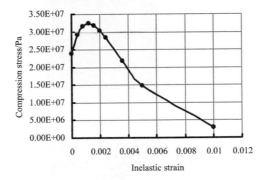

Figure 2.10. Data that defines the plastic strain hardening–softening under compression.

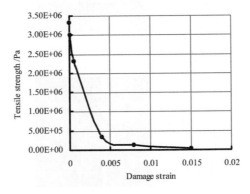

Figure 2.11. Data that defines the plastic softening under tensional loading.

2.3.2.1 *Discretization of the double-notched, four-point shear beam*

Fig. 2.12 shows the mesh of the double-notched, four-point shear beam used in the numerical calculation. The numerical simulation was performed with ABAQUS/Explicit. In this model, the numerical results of damage and strain at time t = 1.25 s, 1.475 s, and 1.4875 s were chosen and analyzed to illustrate the damage initiation and evolution process. A DPZ is the area in which damage values at material points in this region vary near 0 to 1. In the numerical process of loading simulation, time t = 1.25 s is the moment of damage initiation; time t = 1.475 s is the moment of occurrence of a complete damage process zone; and time t = 1.4875 s is the moment when a complete damage process zone moves.

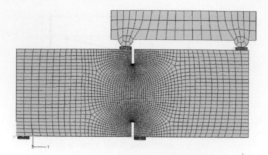

Figure 2.12.   Mesh of the model.

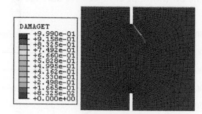

Figure 2.13.   Distribution of tensile damage, t = 1.25 s.

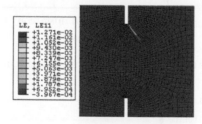

Figure 2.14.   Distribution of tensile strain component, ε_{11}, t = 1.25 s.

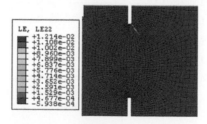

Figure 2.15.   Distribution of tensile strain component, ε_{22}, t = 1.25 s.

Because the loading speed used in the numerical simulation is not identical to that used in the test with the specimen, the time point taken in the numerical calculation does not match the time moment with the same time value, t.

2.3.2.2   *Numerical results obtained with double notched beam*
(1) Distribution of damage and strain at t = 1.25 s.
In Fig. 2.13 through Fig. 2.16, the strain components, ε_{11}, ε_{22}, ε_{12}, and damage variable, D, localize into a narrow band, as shown in the lighter color. There is only a localization band in the structure at the moment.

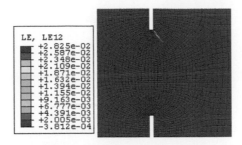

Figure 2.16. Distribution of tensile strain component, ε_{12}, t = 1.25 s.

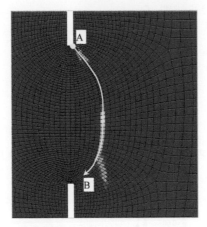

Figure 2.17. Path AB along which band of damage localization develops.

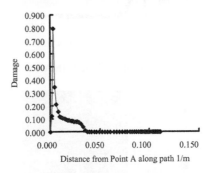

Figure 2.18. Distribution of damage D along Path AB, t = 1.25 s.

Fig. 2.17 shows the Path AB; Fig. 2.18 to Fig. 2.20 show the damage and strain variables along Path AB for time t = 1.25 s. As shown in Fig. 2.18, the maximum value of damage is 0.8; consequently, a complete damage process has not yet formed. A comparison of Fig. 2.19 and Fig. 2.21 shows that the maximum shear strain is 0.0057, and the maximum normal strain is 0.0048, which indicates that the process of deformation is a shear-dominated deformation at the moment t = 1.25 s.

(2) Distribution of damage and strain at t = 1.475 s.
Fig. 2.22 through Fig. 2.25 show the distributions of damage, D, and the strain components within the newly-formed complete damage process zone. There is only one band of localization of damage.

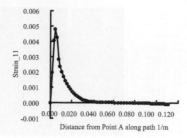

Figure 2.19.   Distribution of strain component ε_{11} along Path AB at t = 1.25 s.

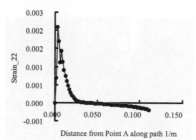

Figure 2.20.   Distribution of strain component ε_{22} along Path AB at t = 1.25 s.

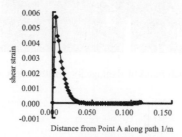

Figure 2.21.   Distribution of strain component ε_{12} along Path AB at t = 1.25 s.

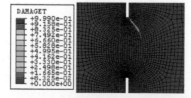

Figure 2.22.   Distribution of damage, D, at t = 1.475 s.

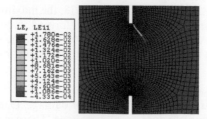

Figure 2.23.   Distribution of strain component, ε_{11}, at t = 1.475 s.

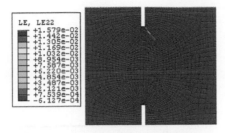

Figure 2.24.   Distribution of strain component, ε_{22}, at t = 1.475 s.

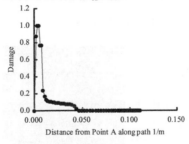

Figure 2.25.   Distribution of strain component, ε_{12}, at t = 1.475 s.

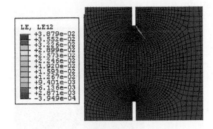

Figure 2.26.   Distribution of damage, D, along Path AB at t = 1.475 s.

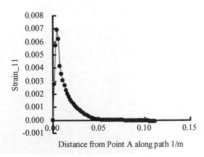

Figure 2.27.   Distribution of strain, ε_{11}, at t = 1.475 s.

Fig. 2.26 through Fig. 2.28 illustrate the distributions of damage, D, and the strain components along Path AB. In Fig. 2.26, the maximum value of damage reaches 1, which indicates an occurrence of a complete damage process zone. When comparing Fig. 2.27 with Fig. 2.28, the value of the maximum shear strain is 0.037, and the value of the maximum normal strain is 0.007 at the end of the damage zone, which indicates that the deformation at the moment is shear-dominated. The front of the damage process zone has 0 damage, as shown in Fig. 2.26. The shear strain in Fig. 2.28 is 0.000088, and the normal strain in Fig. 2.27

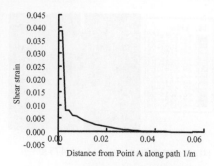

Figure 2.28.   Distribution of strain, ε_{12}, at t = 1.475 s.

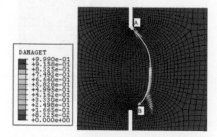

Figure 2.29.   Distribution of damage, D, at t = 1.4875 s.

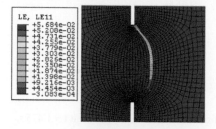

Figure 2.30.   Distribution of strain component, ε_{11}, at t = 1.4875 s.

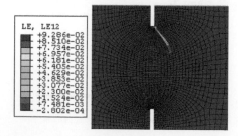

Figure 2.31.   Distribution of strain component, ε_{12}, at t = 1.475 s.

is 0.00095, which is 10 times the shear strain. This result indicates that the deformation at the front point of the damage process zone at the moment is tension-dominated.

(3) Distribution of damage and strain at t = 1.4875
Fig. 2.29 through Fig. 2.31 illustrate the distributions of damage, D, and strain components, ε_{11} and ε_{12}, within the specimen. These figures illustrate the development and movement of the complete damage process zone under a mixed-mode loading condition. Fig. 2.29 shows

that the complete damage process zone develops along Path AB and bifurcates at a point near point B. A secondary damage zone begins to form in the left corner of the lower notch near B, but its appearance is not so obvious yet in Fig. 2.29.

The distribution of damage, D, and strain components at time t = 1.4875 s along Path AB are shown in Fig. 2.32 through Fig. 2.34. By comparing Fig. 2.33 with Fig. 2.34, it is found that the maximum value of shear strain at the tail of the damage process zone is 0.0675, and that of tensile strain is 0.05. This result indicates that the deformation is shear-dominated. Fig. 2.32 and Fig. 2.33 show that at the front of the damage process zone, the maximum value of tensile strain is 0.00065, and the shear strain in Fig. 2.34 at this position is –0.000088, which is less than 1/7 of the tensile strain. It indicates the deformation here is normal strain-dominated.

The length of the damage process zone is the shortest distance between the point with damage value 1 and the point with damage value 0.

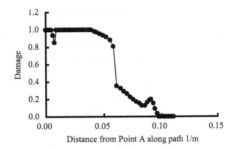

Figure 2.32.   Distribution of damage, D, along Path AB at t = 1.4875 s.

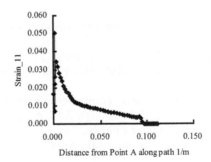

Figure 2.33.   Distribution of strain, ε_{11}, along Path AB at t = 1.4875 s.

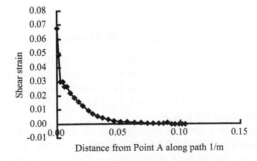

Figure 2.34.   Distribution of strain, ε_{12}, along Path AB at t = 1.4875 s.

Table 2.2.   Values of damage and strain at both ends of a complete DPZ, t = 1.475 s.

t = 1. 475 s	Distance from point A	Damage	ε_{12}	ε_{11}	ε_{22}
At front of DPZ	4.66E-02	0	1.41E-05	9.46E-05	8.81E-06
At end of DPZ	4.57E-03	0.999	7.94E-03	6.96E-03	2.30E-03

Table 2.3.   Values of damage and strain at both ends of a complete DPZ, t = 1.4875 s.

t = 1.4875 s	Distance from point A	Damage	ε_{12}	ε_{11}	ε_{22}
At front of DPZ	9.81E-02	0	−1.13E-04	5.51E-05	−1.13E-04
At end of DPZ	3.69E-02	0.999	4.22E-04	9.51E-03	4.22E-04

Table 2.2 lists the values of the damage and strain components at the front point and end point of the damage process zone at time = 1.475 s. Table 2.3 lists the values of the these variables at time t = 1.4875 s.

With the data listed in Table 2.2, it is derived that the length of damage process zone at t = 1.475 s is 0.0421 m. With the data listed in Table 2.3, it is derived that the length of the damage process zone at t = 1.4875 s is 0.0612 m.

These results indicate that the size of the damage process zone is a variable, which depends on the stress status. With the development of stress status within the area of the damage process zone, the size of the DPZ varies; the size of the DPZ under tension is different from that under shear. Shen and Mroz (2000) have proven analytically that the size of the DPZ for a mode-III anti-plane shear crack for a given load is determined by three factors: stress tensor, material mechanical property (such as strength and Young's modulus), and geometry parameters.

(4) Analysis of the secondary DPZ
In Fig. 2.35, the secondary DPZ appears at the left upper corner of the lower notch at time t = 2.001 s after forming the complete primary DPZ. Fig. 2.36 through Fig. 2.39 provide zoomed views of the domain around the secondary DPZ. These figures visualize the appearance and development of the secondary DPZ, which follows a similar rule to that of the primary DPZ.

The distribution of damage, D, and strain components along Path CD in the domain of the secondary DPZ are shown in Fig. 2.40 through 2.42. At time t = 2.001 s, the complete secondary DPZ appears when the damage value reaches 1 at its tail end.

Table 2.4 lists the values of the damage and strain components at both ends of the secondary DPZ.

Subtracting distance from Point C of Tail of secondary DPZ from that of the Front of secondary DPZ listed in Table 2.4, it is determined that the size of the secondary DPZ is 0.0367 m.

2.3.3   *Numerical results obtained with single-notched beam*

Keeping all the other conditions and parameters the same as in the previous double-notched beam test, this test uses a single-notched beam specimen, which has a length of 0.44 m and a height of 0.1 m. Fig. 2.43 through 2.45 show the numerical results for the distribution of damage, D, and strain components at time t = 1.5 s. Fig. 2.43 shows that a complete DPZ has appeared at this moment along Path A_2B_2.

Fig. 2.46 through Fig. 2.48 show the distribution of damage, D, and strain components along Path A_2B_2. As shown in Fig. 2.47 and Fig. 2.48, the value of normal strain at the tail of the DPZ is 0.00797, and the value of shear strain is 0.0139, which indicates that the deformation at this point is a shear-dominated, mixed-mode deformation. At the front point of

Figure 2.35.   Onset of the secondary DPZ and its location within the specimen.

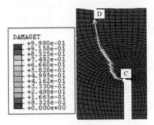

Figure 2.36.   Zoomed view of the secondary DPZ.

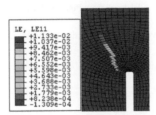

Figure 2.37.   Zoomed view of distribution of strain, ε_{11}, around the DPZ.

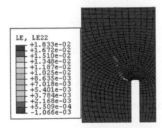

Figure 2.38.   Zoomed view of distribution of strain ε_{22} around the DPZ.

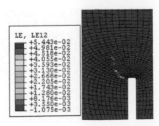

Figure 2.39.   Zoomed view of distribution of strain ε_{12} around the DPZ.

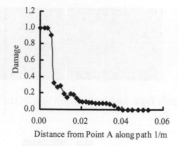

Figure 2.40.   Distribution of damage, D, along Path CD.

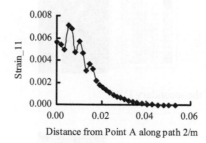

Figure 2.41.   Distribution of strain component ε_{11} along Path CD.

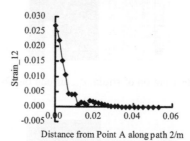

Figure 2.42.   Distribution of strain component, ε_{12}, along Path CD.

Table 2.4.   Values of damage and strain component at both ends of secondary DPZ, t = 2.001 s.

t = 2.001 s	Distance from point C	Damage	ε_{12}	ε_{11}	ε_{22}
Front of secondary DPZ	4.02E-02	0	−7.09E-06	7.45E-05	1.76E-05
Tail of secondary DPZ	3.54E-03	0.999	1.53E-02	4.98E-03	3.84E-03

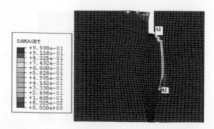

Figure 2.43.   Distribution of damage, D, within the single-notched beam.

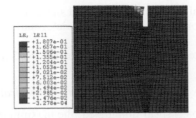

Figure 2.44.   Distribution of strain, ε_{11}, within the single-notched beam.

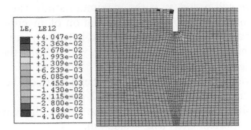

Figure 2.45.   Distribution of strain component, ε_{12}.

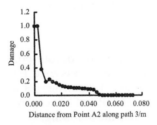

Figure 2.46.   Distribution of damage, D, along Path A2B2.

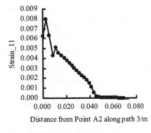

Figure 2.47.   Distribution of strain, ε_{11}, along Path A2B2.

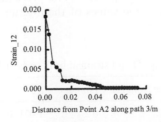

Figure 2.48.   Distribution of strain, ε_{12}, along Path A2B2.

the DPZ, the value of normal strain is 0.000085, and shear strain is 0.00017, which indicates this point is also a shear-dominated, mixed-mode deformation. Table 2.5 lists the values of the variables.

Table 2.5 indicates that size of the DPZ for this test is 0.0473 m.

2.3.4 *Comparisons of the experimental results with the numerical results*

Because the loading condition is shear-dominated, the maximum shear strain is used as the reference variable to calibrate the damage process zone. The damage process zone is determined through two critical strain values: γ_{c1} and γ_{c2}. Parameter γ_{c1} is the minimum strain value below which no damage will occur; γ_{c2} is the maximum strain value above which damage will reach its limit 1, which corresponds to the initiation of macro-crack. A region with a maximum shear-strain value continuously distributed between γ_{c1} and γ_{c2} will be regarded as damage process zone. This point has been numerically verified with the same values of γ_{c1} and γ_{c2} adopted in this chapter by the FEM.

From Table 2.2, it is found that $\gamma_{c2} = 0.00794$ for the end point of the DPZ with damage = 1, and $\gamma_{c1} = 1.41 \times 10^{-5}$ for a front point of the DPZ with damage = 0.

With these critical values of shear strain from Fig. 2.7(b), the length of damage process zone can be determined approximately as $L = 0.054$ m.

Another means of roughly determining the values of γ_{c1} and γ_{c2} is to set these values with reference to the Young's modulus and the cohesion of material of the specimen; this process will result in similar results for these two parameters.

2.3.5 *Remarks*

In this chapter, a device and a numerical model of four-point shear beams are developed and used to calibrate the length of the damage process zone. For the given size of a specimen, the length of the damage process zone was determined experimentally as $L = 54$ mm.

The length and evolution law of the DPZ of the concrete specimen were studied numerically. The conclusions obtained include the following:

- The length of a complete DPZ is in the range of 0.0421 m to 0.0612 m for the given specimen of double-notched beam with a height of 0.15 m and length of 0.4 m. The variation of the DPZ length results from changes of the stress status at points within the DPZ. As the distance increases for a point from the notch, the stress status varies from a shear-dominated status to a tension-dominated status.
- A secondary DPZ will appear after the complete primary DPZ occurs. For the given specimen and loading, the length of the secondary DPZ is 0.0367 m.
- For the single-notched beam described in this chapter, the length of its DPZ is 0.0473 m.

This chapter provides comparisons between the numerical and experimental results and the comparisons are in good accordance. Trends of localization of the strain and damage obtained from experiments are similar to those obtained from the numerical results; the primary DPZ appears first, followed by the secondary DPZ.

The result of the length of damage process zone presented in this chapter provides an experimental basis for determining the values of the internal length, which is essential in a gradient-dependent damage model.

Table 2.5. Values of damage D and strain at both ends of DPZ.

Time t = 1.5 s	Distance from A2	Damage	ε_{12}	ε_{11}	ε_{22}
At front of DPZ	4.98E-02	0	1.07E-04	8.51E-05	4.70E-03
At tail of DPZ	2.52E-03	0.999	1.39E-02	7.97E-03	−1.01E-04

ACKNOWLEDGEMENTS

Partial financial support from China National Natural Science Foundation through contract 10872134, support from Liaoning Provincial Government through contract RC2008-125 and 2008RC38, support from the Ministry of State Education of China through contract 208027, and support from Shenyang Municipal Government's Project are gratefully acknowledged.

NOMENCLATURE

$\bar{\varepsilon}_t^{pl}$	= Equivalent plastic strain
σ_{t0}	= Initial tensile strength, Pa
σ_{c0}	= Initial compressive strength, Pa
σ_{c0}	= Ultimate peak compressive strength, Pa
d_t	= Tensile damage
d_c	= Compressive damage
d	= Synthetic damage variable
E_0	= Young's modulus of the initial intact material, Pa
σ_t	= Stress for tension, Pa
σ_c	= Stress for compression, Pa
$\bar{\sigma}_t$	= Effective stress for tension, Pa
$\bar{\sigma}_c$	= Effective stress for compression, Pa
E	= Young's modulus for damaged material
s	= Stress state index
s_t	= Function of stress state for tension
s_c	= Function of stress state for compression
w_t	= Weight parameter for tension
w_c	= Weight parameter for compression
$\mathbf{D}_0^{el}$	= Matrix of stiffness of the intact materials, Pa
$\hat{\sigma}_i$,	
$(i=1,2,3)$	= Principal stress components of effective stress tensor
G	= Plastic potential, Pa
$\psi(\theta, f_i)$	= Dilatancy angle
$\sigma_{t0}(\theta, f_i)$	= Threshold value of the tensile stress at which damage initiates, Pa
$\varepsilon(\theta, f_i)$	= Model parameter that defines the eccentricity of the loading surface in the effective stress space
$\hat{\bar{\sigma}}_{max}$	= Value of maximum principal stress component, Pa
σ_{b0}/σ_{c0}	= Ratio between the limit of bi-axial compression and that of uniaxial compression
k_c	= Ratio between the second invariant of the stress tensor at the tensile meridian $q(TM)$ and the second invariant at the Compressive meridian $q(CM)$ under arbitrary pressure
$q(TM)$	= Second invariant of the stress tensor at the tensile meridian, Pa
$q(CM)$	= Ssecond invariant at the compressive meridian, Pa
$\bar{\sigma}_t(\bar{\varepsilon}_t^{pl})$	= Effective tensile strength, Pa
$\bar{\sigma}_c(\bar{\varepsilon}_c^{pl})$	= Effective compressive strength, Pa
γ_{max}	= Maximum shear strain
DPZ	= Damage process zone

REFERENCES

Aifantis, E.C.: On the role of gradient in the localization of deformation and fracture. *Int. J. Engng. Sci.* 30 (1992), pp. 1279–1299.

Aifantis, E.C.: Update on a class of gradient theories. *Mech. of Mater.* 35:1(2003), pp. 259–280.

Bazant, Z.P. and Cedolin, L.: S*tability of Structures: Elastic, Inelastic, Fracture And Damage Theories.* Oxford University Press, Oxford, New York, 1991.

Bazant, Z.P. and Pijaudier-Cabot, G.: Nonlocal continuum damage, localization instability and convergence. *ASME J. Appl. Mech.* 55 (1988), pp. 287–293.

Chaboche, J.L: Damage induced anisotropy: on the difficulties associated with the active/passive unilateral condition. *Int. J. Dama. Mech. 1* (1992), pp. 148–171.

de Borst, R., Pamin, J., and Geers, M.G.D.: On coupled gradient-dependent plasticity and damage theories with a view to localization analysis. *Eur. J. Mech. A/Solids* 18 (1999), pp. 939–962.

Govindjee, S., Kay, G. and Simo, J.C.: Anisotropic modelling and numerical simulation of brittle damage in concrete. *Int. J. Numer. Meth. Engng*, 38 (1995), pp. 3611–3633.

Lee, J. and Fenves, G.L.: Plastic-damage model for cyclic loading of concrete structures. *ASCE J. Engng Mech. 124*:8 (1998), pp. 892–900.

Lemaitre, J: *A Course on Damage Mechanics.* 2nd ed., Berlin: Springer, 1990.

Lubliner, J., Oliver, J., Oller, S. and Onate, E.: A plastic damage model for concrete. *Int. J. Solids & Struct.* 25:3 (1989), pp. 299–326.

Mazars, J. and Pijaudier-Cabot, G.: Continuum damage theory – application to concrete. *ASCE J. Engng. Mech.* 115:2 (1989), pp. 346–365.

Meschke, G., Lackner, R. and Mang, H.A.: An anisotropic elastoplastic-damage model for plain concrete. *Int. J. Numer. Meth. Engng.* 42 (1998), pp. 703–727.

Nechnech, W., Meftah, F. and Reynouard, J.M.: An elasto-plastic damage model for plain concrete subjected to high temperatures. *Engng. Strut.* 24:5 (2002), pp. 597–611.

Saanouni, K., Chaboche, J.L. and Lesne, P.M.: Creep crack growth prediction by a non-local damage formulation. In: Mazars, J. and Bazant, Z.P. (eds): *Proceedings of Europe-US Workshop on Strain Localization and Size Effects in Cracking and Damage*, Cachan, 6–9 September 1988, Elsevier Applied Science, London and New York, 1988, pp. 404–414.

Shen, X. and Mroz, Z.: Shear beam model for interface failure under antiplane shear (I)-fundamental behavior. *Applied Mathematics and Mechanics: English Edition* 21:11 (2000), pp. 1221–1228.

Shen, X.P., Shen, G.X., Chen, L.X. and Yang, L.: Investigation on gradient-dependent nonlocal constitutive models for elasto-plasticity coupled with damage. *Applied Mathematics and Mechanics-English Edition* 26:2 (2005), pp. 218–234.

Swoboda, G., Shen, X.P. and Rosas, L.: Damage model for jointed rock mass and its application to tunneling. *Comput. Geotechn.* 22:3/4 (1998), pp. 183–203.

CHAPTER 3

Trajectory optimization for offshore wells and numerical prediction of casing failure due to production-induced compaction

Xinpu Shen, Mao Bai, William Standifird & Robert Mitchell

3.1 INTRODUCTION

Trajectory optimization is a fundamental aspect of a wellbore design. A deliberately optimized wellbore trajectory enables drilling to be performed under minimum geostress loads and promotes a longer service life for casings. Trajectory optimization is particularly significant to projects in which wellbores are designed with reference to a given platform. Although platform drilling has historically been an offshore consideration, an increasing number of field development designs include multiple wellbores drilled from a single surface location. Consequently, the necessity for trajectory optimization increases with the constraint of a fixed surface location to an irregular reservoir geometry.

Despite the importance of trajectory optimization, very few reference papers are available on this subject. In this work, numerical methods are adopted to evaluate the advantages of each possible choice of wellbore trajectory. A key issue is the choice of mechanical variables as an index. According to key references (e.g., Sulak and Danielsen 1989; Sulak 1991), the principal types of casing failure in abnormally high geostress zones and/or in cases in which the casing was subjected to production-induced stresses can be classified as follows: compressive inward collapse, shear failure, axial plastic compressive buckling, and tensile plastic failure in the axial direction.

The data used in this chapter is provided as illustrative data only; it is not the real data of any commercial project, but includes modified data that is based on real cases. Accuracy and completeness of models used in this chapter are validated and ensured.

3.2 GEOTECHNICAL CASING DESIGN AND OPTIMAL TRAJECTORIES

Casing and tubing strings are the main parts of the well construction. All wells drilled to produce oil/gas (or to inject materials into underground formations) must be cased with material of sufficient strength and functionality. Casing is the major structural component of a well. It is needed to maintain borehole stability, prevent contamination of water sands, isolate water from producing formations, and control well pressures during drilling, production, and workover operations. Casing provides locations for the installation of blowout preventers, wellhead equipment, production packers, and production tubing. The cost of casing is a major part of the overall well cost; consequently, the selection of casing size and grade, connectors, and setting depth is a primary engineering and economic consideration.

The fundamental basis of casing design is that if stresses in the pipe wall exceed the yield strength of the material, a failure condition exists. Consequently, the yield strength is a measure of the maximum allowable stress. To evaluate the pipe strength under combined loading conditions, the uniaxial yield strength is compared to the yielding condition. Perhaps the most widely accepted yielding criterion is based on the maximum distortion energy theory, which is known as Huber-Hencky-Mises yield condition, or as the von Mises stress, "triaxial stress," or equivalent stress (Crandall and Dahl 1959). "Triaxial stress" (equivalent stress) is not a true stress. It is a theoretical value that enables a generalized three-dimensional stress state to be compared with a uniaxial failure criterion (the yield strength). In other words, if the triaxial stress exceeds the yield strength, a yield failure is indicated. The triaxial safety factor is the ratio of the material's yield strength to the triaxial stress.

The yielding criterion is stated as follows:

$$\sigma_{VME} = \frac{1}{\sqrt{2}} \sqrt{\left(\sigma_z - \sigma_\theta\right)^2 + \left(\sigma_\theta - \sigma_r\right)^2 + \left(\sigma_r - \sigma_z\right)^2} \geq Y_p \qquad (3.1)$$

where Y_p is the minimum yield strength, σ_{VME} is the von Mises equivalent stress, σ_z is axial stress, σ_θ is tangential or hoop stress, and σ_r is the radial stress.

Although it is acknowledged that the von Mises criterion is the most accurate method of representing elastic yield behavior, the use of this criterion in tubular design should be accompanied by these precautions:

• For most pipe used in oilfield applications, collapse is frequently an instability failure that occurs before the computed maximum triaxial stress reaches the yield strength. Consequently, triaxial stress should not be used as a collapse criterion. Yielding occurs before collapse only in thick-wall pipe.
• The accuracy of the triaxial analysis depends upon the accurate representation of the conditions that exist both for the pipe as installed in the well and for the subsequent loads of interest. Often, it is the *change* in load conditions that is most important in stress analyses. Consequently, an accurate knowledge of all *temperatures* and *pressures* that occur over the life of the well can be critical to accurate triaxial analyses.

The following references are recommended for a more in-depth review of casing design practices: Mitchell 2006; Economides *et al.*, 1998; Aadnoy 1996; Rabia 1987; and American Petroleum Institute 1983 and 1983.

The literature focused on geotectonic loading is also extensive. The following is, at best, a representative sampling of casing design in subsiding formations: Rieke and Chilingarian 1974; Mitchell and Goodman 1977; Smith and Clegg 1971; Smith *et al.*, 1973; Weiner *et al.*, 1975; Bruno 1992; Bickley and Curry 1992; Roulffignac *et al.*, 1995; Dale *et al.*, 1996; Hilbert *et al.*, 1998; Bruno 2001; Ibekwe *et al.*, 2003; Li *et al.*, 2003; and Sayers *et al.*, 2006.

It is clear that casing design is influenced by several factors and scenarios. To simplify the optimization process described in this chapter, the description focuses only on the incremental loading arising from formation displacement and compaction. This load represents a loading in addition to the other loads and conditions of conventional casing design. Therefore, the von Mises stress used as the optimization criterion is not the actual von Mises stress of the casing because we have not dealt with the casing running, cementing, and the specific pressure and temperature loads normally considered in casing design. Rather, it has been assumed that the loads from subsidence are the primary loads that the casing must accommodate, and that the other loads are of secondary importance. After a suitable trajectory has been determined, this worst-case assumption can be tested by performing a thorough conventional casing analysis.

Because of the large differences between field scale and reservoir scale modeling, it has been difficult, if not impossible, to combine these models in the past. In fact, existing examples of numerical analyses of casing failure were either performed at the reservoir scale without directly linking to behaviors at the field scale, or performed at a much larger scale, which sacrificed much needed modeling resolution. Capasso and Mantica (2006) reported the evaluation of reservoir compaction and surface subsidence induced by hydrocarbon production by using a one-way coupled model for porous flow linked with elasto-plastic deformation. Bruno (1992) reported casing failure caused by reservoir compaction in a semi-analytical manner.

The task of this work is to select an optimized well trajectory with two given ends at points A and B, shown in Figure 3.1. This optimized well trajectory should present the least resistance to drilling and ensure that the casing system can survive changes in pore pressure and in-situ stress caused by near- and far-field changes attributable to production. Abaqus submodeling techniques are used to manage the field to reservoir scale discrepancy. The concept of the

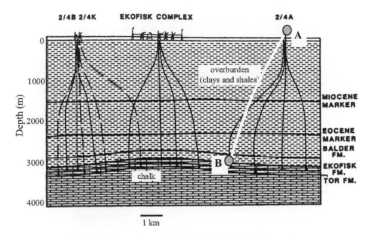

Figure 3.1.   Geostructure and wellbore distribution in the area of Ekofisk (modified after Sulak and Danielsen 1989).

submodeling technique includes using a large scale global model to produce boundary conditions for a smaller scale submodel. In this way, the hierarchical levels of the submodel are not limited. Using this approach, a highly inclusive field scale analysis can be linked to a very detailed casing stress analysis at a much smaller scale. The benefits are bidirectional, with both the larger and smaller scale simulations benefiting from the linkage.

Because of the complex geology and non-uniform distribution of petroleum, casing failure has been a common incident at Ekofisk (Sulak and Danielsen 1989; Sulak 1991). As a result of the casing failures and the pursuit of a solution, the Ekofisk field in the North Sea has been investigated by various researchers since the 1970s. Initially, Finol and Farouq Ali (1975) developed a two-phase 2D computer program for the prediction of the reservoir behavior. Later, Yudovich *et al.* (1989) provided a detailed description of the casing integrity challenges in the Ekofisk field, including observations of both casing compaction and tension. More recently, Lewis *et al.* (2003) used a 3D calculation of the Ekofisk field that used porous fluid flow and elastic deformation.

The following sections of this chapter first present the geometry of the field scale model of the Ekofisk field with constitutive models for a chalk reservoir. Second, the numerical results of a field scale model obtained through numerical modeling are presented. The calculations presented adopt a multi-physics model in which the visco-plastic deformation is combined with the porous fluid flow. The boundary conditions of the first submodel are taken directly from the numerical results of the field scale model. A finer mesh is used in the secondary submodel, and the numerical analysis at the casing section scale is performed by using the results of first submodel as boundary conditions. This chapter also presents an elasto-plastic analysis of the casing stressed by subsidence and pressure depletion, followed by the conclusions from the modeling and analysis.

3.3   THE WORK PROCEDURE

Figure 3.2 shows a three-step workflow to optimize a wellbore trajectory between a drilling platform and a given reservoir target. These steps include the following:

- Construct a field scale model, including a visco-elasto-plastic deformation analysis and the porous fluid flow related to pressure depletion. Plot the distributions of the von Mises stress along each wellpath candidate. The path along which its maximum von Mises stress value is the minimum among those of the three paths will be chosen as the best path of trajectory.

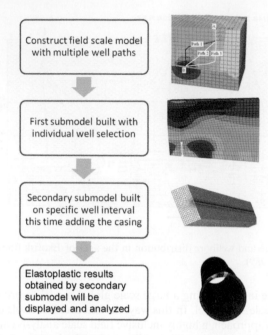

Figure 3.2.   Flowchart for using submodeling techniques for trajectory optimization.

- Construct a primary submodel using the wellbore path with the least von Mises stress. In this submodel, the hydropressure effects on the wellbore surface will be included. The distribution of the von Mises stress along the wellbore trajectory will be plotted and analyzed. The wellbore interval in which the maximum von Mises stress occurs will be included in the secondary submodel.
- Construct a secondary model of the previously selected wellbore interval, this time including the specified tubular that will line the wellbore. In this model, an elasto-plastic prediction of casing failure will be made.

3.4   THE MODEL

3.4.1   *Model geometry*

The field scale model is shown in Figure 3.3. The total depth of the model is 4000 m, the width is 5500 m, and the length is 9000 m; the distribution of the chalk reservoir is shown in red. The model uses four vertical layers of overburden; the first layer is 1500 m, the second is 800 m, the third is between 435 and 800 m, and the bottom layer is between 900 and 1265 m. The reservoir layer that ranges from 50 to 150 m is located in the lower middle of layer 3, as shown in Figure 3.3.

As shown in Figure 3.1, the horizontal distance between the end points of two reservoir intersections is approximately 2000 m. This distance suggests that the radial displacement from each wellbore, where the effect of pressure depletion would be expected to be encountered, is approximately 1000 m. Consequently, the local pressure depletion around a wellbore is assumed to have a circular area of influence, as shown in Figure 3.4. The horizontal distance between points A and B, shown in Figure 3.1, is 2100 m; point B is located in the center of the circle area of Figure 3.4.

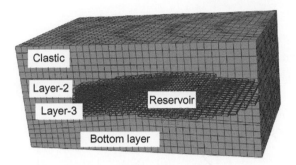

Figure 3.3.   Distribution of reservoir.

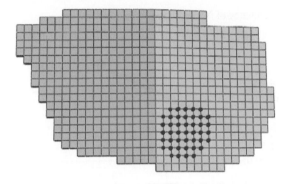

Figure 3.4.   Distribution of first influence area of wellbore concerned (marked with red dots).

3.4.2   *Material models*

The Ekofisk chalk is complex; this complexity creates issues related to visco-plasticity (Hickman 2004) and to compatibility (Cipolla *et al.,* 2007). Furthermore, the chalk elastic modulus varies with pressure in effective stress space.

This work adopts the modified Drucker-Prager yielding criterion. The cohesive strength and frictional angle are given the following values: $c = 1$ MPa, $\phi = 25°$.

The creep law, given in the following equation (Abaqus Manual 2008), is adopted:

$$\dot{\bar{\varepsilon}}^{cr} = A\left(\bar{\sigma}^{cr}\right)^{n} t^{m} \tag{3.2}$$

where $\dot{\bar{\varepsilon}}^{cr}$ represents the equivalent creep strain rate; $\bar{\sigma}^{cr}$ represents the von Mises equivalent stress; t is the total time variable, A, n, m are three model parameters that are given the following values:

$$A = 10^{-21.8}, \quad n = 2.667, \quad m = -0.2 \tag{3.3}$$

The compaction property of the chalk reservoir is simulated with a linear law of hardening. Chalk skeleton variations of both Young's modulus and Poisson's ratio with pressure in the effective stress space are expressed in Figure 3.5 and Figure 3.6, respectively.

The property of pressure dependency of chalk is realized by using the Abaqus user subroutine USDFLD in the calculation. The porosity parameters of chalk are given the

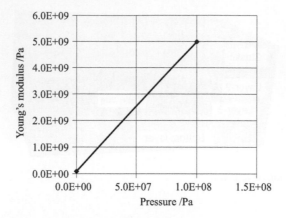

Figure 3.5.   Pressure dependency of Young's modulus.

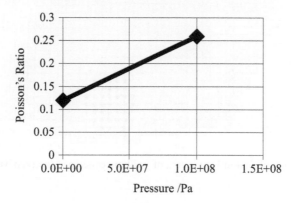

Figure 3.6.   Pressure dependency of Poisson's ratio.

following values: initial void ratio $R = 0.5$; intrinsic permeability coefficient $k = 2$ Darcy. The following brief notes are necessary for using the USDFLD subroutine:

- The user subroutine USDFLD is typically used when complex material behavior must be modeled and the modeler does not want to develop a UMAT subroutine.
- Most material properties in Abaqus can be defined as functions of field variables, f_i.
- The subroutine USDFLD enables the modeler to define f_i at every integration point of an element.
- The subroutine has access to solution data. Therefore, the material properties can be a function of the solution data.
- The subroutine USDFLD can be used only with elements that require a material definition.
- Typically, the modeler must define the dependence of the material properties, such as elastic modulus or yield stress, as functions of field variables, f_i.
- The pressure dependency can be accomplished using either tabular input or additional user subroutines.

The source file of the user subroutine is shown below as:

```
c
    SUBROUTINE USDFLD(FIELD,STATEV,PNEWDT,DIRECT,T,CELENT,
   1 TIME,DTIME,CMNAME,ORNAME,NFIELD,NSTATV,NOEL,NPT,LAYER,
   2 KSPT,KSTEP,KINC,NDI,NSHR,COORD,JMAC,JMATYP,MATLAYO,LACCFLA)
```

```
C
   INCLUDE 'ABA_PARAM.INC'
C
   CHARACTER*80 CMNAME,ORNAME
   CHARACTER*3 FLGRAY(15)
   DIMENSION FIELD(NFIELD),STATEV(NSTATV),DIRECT(3,3),
 1 T(3,3),TIME(2),sTRESS(6)
   DIMENSION ARRAY(15),JARRAY(15),JMAC(*),JMATYP(*),COORD(*)
C
     CALL SINV(STRESS,SINV1,SINV2,NDI,NSHR)
     FIELD(1) = SINV1/3.d0
C
       STATEV(1) = FIELD(1)
C
   RETURN
     END
```

A utility that subroutes SINV has been called in the USDFLD. Its function is to calculate the invariants of a stress tensor represented by STRESS. SINV1 represents the first stress invariant, and it is 1/3 is the so-called mean stress; it is also known as 'pressure.' This variable will be recorded as the data array STATE(1) and be referred when the program calculates the elastic stiffness matrix in terms of Young's modulus and Poisson's ratio.

The input data, which connects the pressure-dependant behavior of elasticity to the calculation, is shown below with underlined italics:

```
**
** MATERIALS
**
*Material, name=RESERVIOR
*Depvar
    1,
*Elastic, dependencies=1
 5.e9,   0.26,   -100e6.
 1e+08,  0.12,   -10000.
*User Defined Field
*Permeability, specific=0.2
 1e-12, 0.1
 1e-10, 0.6
*Drucker Prager
 28.9, 1., 28.9
*Drucker Prager Creep, law=TIME
 2.5e-23, 2.942, -0.2
*Drucker Prager Hardening, type=SHEAR
 1.2e+07,0.
*Density
 2200.,
```

The clastic layer on the top of the model and the bottom layer material of the model are assumed to be elastic. Layer 2 and layer 3 materials are assumed to be visco-elasto-plastic.

3.4.3   *Loads and boundary conditions of the global model*

The depth of the overburden seawater is 100 m. The seawater produces a uniform pressure of 0.98 MPa on the overburden rock of the field scale model. The geostress field is balanced by the gravity field in the vertical direction, and components of lateral stress are assumed to have a value of 90% of the vertical component. The density values of the reservoir and the four model layers are given as:

$$\rho_{reservoir} = 2100 \ kg/m^3, \rho_{clatic} = 2200 \ kg/m^3, \rho_{layer-2} = 2250 \ kg/m^3$$
$$\rho_{layer-3} = 2250 \ kg/m^3, \rho_{bottom} = 2500 \ kg/m^3$$

(3.4)

The initial pore pressure within the reservoir is assumed to be 34 MPa.

The purpose of the field scale model calculations is to estimate the distribution of the von Mises equivalent stress and its variation with pressure depletion. This information is then used to select the optimal trajectory based on the stresses to which the casing will be subjected. Therefore, in this field scale model, no actual wellbore exists, only possible trajectories. The effects of pore fluid pressure and the redistribution of in-situ stress around a wellbore will be considered in the submodel, and are neglected in the field scale models.

In the calculation, a local pore pressure depletion of 34 to 10 MPa will occur first to simulate the subsidence caused by production of the well studied. Subsequently, a second field scale pore pressure depletion of 34 to 20 MPa will occur to simulate the influence of nearby production wells.

3.5   NUMERICAL RESULTS OF THE GLOBAL MODEL

Figure 3.7 shows the results of subsidence after pressure depletion in the area surrounding the studied wellbore. The top of the block is at the depth of 2566 m, just above the chalk reservoir. The maximum subsidence above the reservoir is approximately 6 m and is consistent with key references (Sulak and Danielsen 1989; Sulak 1991). Figure 3.8 shows the subsidence results after further pressure depletion of the entire Ekofisk field. These subsidence results indicate that the assumed values of the model parameters are reasonable with reference to existing subsidence observations.

Figure 3.9 shows three candidate well paths for further optimization between points A and B. Figure 3.10 shows the numerical results of the distribution of the von Mises equivalent stress along the three paths, and Figure 3.11 shows the numerical results of the distribution of the von Mises equivalent stress after pressure depletion on the field scale. Figure 3.10 shows that the maximum value of the von Mises equivalent stress along Path-1 is the least of the three candidates. Furthermore, Figure 3.11 shows that this maximum value decreases as the pore pressure outside of the local region decreases. Because the von Mises equivalent stress is a potential index of distortion deformation, the distortion deformation situation will be improved with the development of wellbores in the nearby field. These two phenomena suggest that Path-1 is the optimal path and will result in the minimum potential stress loading on the casing.

Based on these observations, a submodel is used for additional analysis of Path-1.

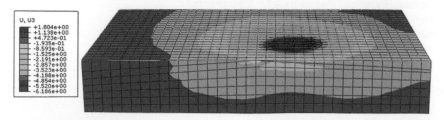

Figure 3.7.   Subsidence field after pressure depletion near the wellbore (3D sectional view at depth 2566 m).

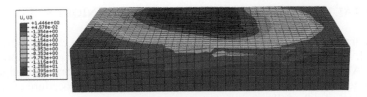

Figure 3.8.   Subsidence field after pressure depletion occurred in the entire field (3D sectional view at depth 2566 m).

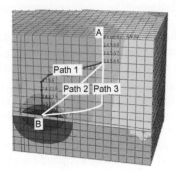

Figure 3.9.   The paths that represent possible wellbore trajectories (sectional view).

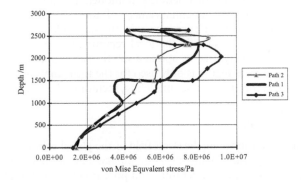

Figure 3.10.   Comparison of the von Mises stress along the three paths shown in Figure 3.9 after pressure depletion near the wellbore.

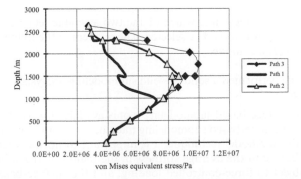

Figure 3.11.   Comparison of the von Mises stress along the three paths shown in Figure 3.9 after pressure depletion of the entire field.

3.6   GENERAL PRINCIPLE OF SUBMODELING TECHNIQUES

Submodeling is the study of a local section of a model based on an existing solution from a global model. A detailed description of the principle and procedure of submodeling can be found in the Abaqus User's Manual (2010). This chapter presents a brief introduction of the submodeling technique for the convenience of readers. The basic procedure of analysis with submodeling includes the following:

- Obtain a global solution using a coarse mesh of global model.
- Interpolate this solution onto the boundary of a locally refined mesh of the submodel. Typically, a displacement solution will be used to interpret onto the submodel boundary in structural analysis; sometimes, however, it is limiting to use a stress solution for the interpretation of boundary of the submodel. Other degrees of freedom may also be used for this purpose, such as nodal rotations, nodal temperature, and pore pressure.
- Obtain a detailed solution in the local area of interest with a refined mesh of the submodel.

Assumptions adopted in submodeling technique include the Saint-Venant's principle, which indicates that the boundary of the submodel is sufficiently far from the region within the submodel where the response changes. Therefore, the global model solution defines the response on the submodel boundary.

It is also assumed that detailed modeling of the local region has a negligible effect on the global solution. It is also required that the loads applied on submodel should be the same as those applied on the global model.

An initial global analysis of a structure usually identifies areas where the response to the loading is deemed to be crucial. Further detailed submodeling will focus on this crucial area. Submodeling provides an easy model enhancement of these areas without to the need to remesh and reanalyze the entire model; consequently, the submodeling process reduces analysis costs and provides detailed results in the local regions of interest. During the analysis process, a coarse model is usually used to obtain a solution around a local region of interest, and a refined mesh of the local region is only used for the local analysis.

Abaqus provides two techniques for submodeling, including the following:

- Node-based submodeling;
- Surface-based submodeling.

For the surface-based submodeling, the stress field of the global solution is interpolated onto the submodel surface integration points at its boundary. For the node-based submodeling, the displacement field and/or the temperature and pore pressure of the global solution will be interpolated onto the submodel nodes at its boundary.

Of these two techniques, node-based submodeling is more general and more common than surface-based submodeling. Either technique or a combination of both can be used in an analysis.

The questions may arise regarding which of the two available techniques would provide the best fit, and which option should be selected. The answers to the following questions can be used to select the best technique:

- Is the analysis to be performed a solid-to-solid submodeling analysis?
- Is it a static analysis?

The technique for surface-based submodeling is only available for solid-to-solid and static analyses. Consequently, node-based submodeling must be used for all other procedures.

When selecting the submodel technique, it is important to notice whether or not there is a significant difference in the average stiffness in the region of the submodel. If so, the global model is subject to force-controlled loading, and the surface-based technique will generally yield more accurate stress results than those obtained with the node-based technique. Otherwise, when the stiffness is comparable, node-based submodeling will provide similar results to surface-based, but with less risk of numerical issues (caused by rigid-body modes).

Differences in the stiffness of the global and submodels may arise from additional details in the submodel (such as a fillet or a hole), or from minor geometric changes that do not warrant re-running the global analysis.

The workflow for submodeling is listed as follows:

- Define the problem. Ensure that the Saint-Venant's principle is valid for the submodel. Carefully select appropriate submodeling boundaries.
- Run the global model, ensuring that the appropriate output has been selected to drive the submodel.
- Check the global model results, particularly in the areas that will be used to drive the submodel. These areas should be free from local irregularities.
- Define the submodel.
- Apply the loads to the submodel in accordance with the loading in the global model.
- Apply the submodel boundary conditions (node-based) or the surface boundary tractions (surface-based).
- Apply inertia relief, if required.
- Run the submodel analysis.
- Verify submodeling results.

In the following sections, node-based submodeling techniques will be used.

3.7 FIRST SUBMODEL

Using Path-1 as the trajectory with the lowest casing stress given the surface and reservoir position constraints, the first submodel is built, as shown in Figure 3.12. The region addressed by this submodel is much smaller than the field scale modeling; only the lower-right corner of the field scale model (Figure 3.4) is included in first submodel. Vertically, the depth of the submodel is adjusted to just above the reservoir. For simplicity, the deformation and the porous flow are only calculated in the field scale model. Field scale calculations provide the boundary conditions to this first submodel, which can accurately account for the influence of pressure depletion within the reservoir.

The loads on the first submodel include the following:

- In-situ stress field generated by gravitational loading
- Vertical stress created by the seawater load
- Hydraulic pressure applied on wellbore surface

The wellbore is built into the first submodel along Path-1.

The boundary conditions are set by applying the displacement constraints, obtained from numerical results of the field scale model, on the four lateral sides and bottom of the submodel. Because the reservoir is not included in this submodel, the calculation involves only visco-elasto-plastic static deformation; no porous fluid flow is considered.

The input data, which connects the submodel with global model, is shown below as:

```
** BOUNDARY CONDITIONS
** Name: BC-1 Type: Submodel
*Boundary, submodel, step=1, timescale
_PickedSet104, 1, 3
```

where the nodal set "_PickedSet104" is defined manually in the process of CAE modeling at the six outer surfaces of the submodel.

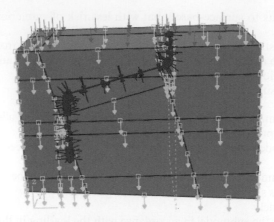

Figure 3.12.   Illustration of loads on the first submodel.

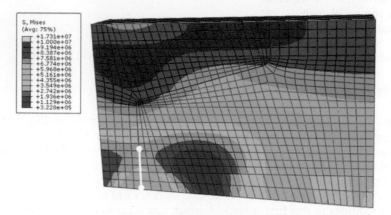

Figure 3.13.   Distribution of the von Mises stress within local model. The darker area represents the area
in which the von Mises stress value is greater (sectional view along trajectory, unit: in Pa).

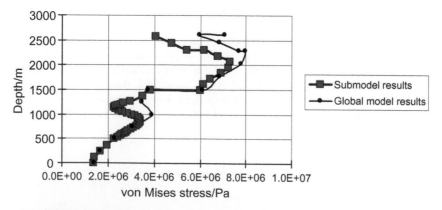

Figure 3.14.   Distributions of the von Mises stress (thick line) and its comparison with the one obtained
in global model (thin line) along the wellbore trajectory.

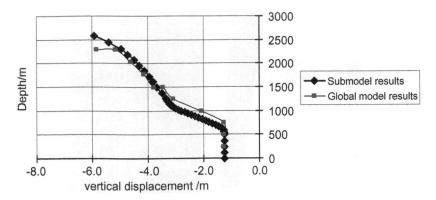

Figure 3.15.   Distribution of vertical displacement component (thick line) and its comparison with the one obtained in global model (thin line) along the wellbore trajectory.

3.7.1   *Local model results*

Figure 3.13 shows the distribution of the von Mises equivalent stress in a section for Path-1. Figure 3.14 illustrates a comparison of the distribution of the von Mises equivalent stress along Path-1 obtained by the field scale model and that of the first submodel. The values of the local result of the von Mises stress are less than the values obtained by the field scale model at several points. For the purpose of further understanding the submodel results, a comparison between subsidence results obtained by the field scale and submodel are made and displayed in Figure 3.15. The two sets of results are in close agreement, although the submodel results will be more accurate because of the higher resolution.

As shown in Figure 3.13 and Figure 3.14, the greatest von Mises stress values occur in a region 400 m above the reservoir, along the wellbore trajectory. This finding indicates that this location has the greatest potential for casing distortion. Consequently, a secondary submodel and more detailed analysis, which will include casing, will be performed for this region. This analysis will ensure that the tubular selected will endure the stresses convolved on the selected trajectory wellbore.

3.8   SECONDARY SUBMODEL AND CASING INTEGRITY ESTIMATE

A secondary submodel is used to further refine the mesh in the length of depth indicated by the white line at the left lower corner of Figure 3.13. Casing is set along the entire length of well trajectory, with an internal diameter of 0.254 m (10 in.) and a wall thickness of 0.015 m (approximately 3/5 in.). The casing material is assumed to be elasto-plastic with the following values of elastic and strength parameters:

$$E = 2 \times 10^{11}\,\text{Pa}, \quad \nu = 0.3, \quad \sigma_s = 8 \times 10^8\,\text{Pa} \tag{3.5}$$

Figure 3.16 and Figure 3.17 show the results obtained with the secondary submodel. The distribution of the von Mises stress within the rocks is shown in Figure 3.16; the right side of Figure 3.16 shows a zoomed view of the upper end of the secondary local model.

Figure 3.17 shows the distribution of plastic strain within the casing. Plastic deformation occurs at a small portion of casing at its right end (upper end as z-axis upward). The maximum value of plastic strain is 0.0095. Although this value is greater than a standard initial plastic strain value of 0.002 for steel material (Gere and Timoshenko 1987), it is less than the tolerable failure strain value 3.5% (i.e., 0.035) for ductile casing steel (Kaiser 2009). Because

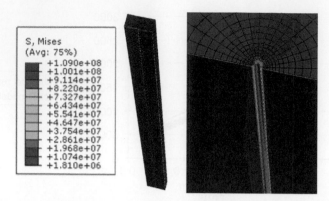

Figure 3.16.   Distribution of von Mises stress within rock of secondary submodel (sectional view along trajectory, unit: in Pa).

Figure 3.17.   Distribution of plastic strain within casing.

the field scale model (including geometry and loads) is not symmetrical, the deformation of the casing is also not symmetrical.

3.9   CONCLUSIONS

The optimization of a well trajectory between a surface platform and reservoir intersection in the Ekofisk field was performed. Individual analyses at a casing section scale and an analysis at the field scale were deliberately separated to overcome scale incompatibility and to improve the calculation accuracy. Submodeling techniques were adopted to link the field and reservoir/casing scale challenges and to improve the overall effectiveness of the well path optimization. Subsidence was simulated at the field scale, whereas casing failure was calculated at a local level. Inelastic-visco deformation of the reservoir and porous fluid flow were calculated in the field scale modeling and linked, through submodeling, to the local level.

An index of the von Mises equivalent stress within the rock under various loading conditions, such as pressure depletion and gravity, was used to derive a preferred well path candidate from three different trajectories. This study validates that an optimized trajectory can be achieved if the Path-1 well trajectory is selected because it results in the minimum distortion deformation of the casing.

The proposed numerical procedure provides an effective tool for selecting an optimized trajectory for efficient drilling, and for maximized casing and wellbore stability. General economics will be improved with the reduction in non-productive time, reduced drilling cost, and improved reservoir production, as a result of the enhanced well stability.

NOMENCLATURE

A	=	Creep model parameter
c	=	Cohesive strength, F/L^2, Pa
E	=	Young's modulus, F/L^2, Pa
k	=	Intrinsic permeability coefficient, Darcy, d
m	=	Creep model parameter
n	=	Creep model parameter
R	=	Initial void ratio
t	=	Total time variable, s
ρ_{clatic}	=	Density of clastic, m/L^3, kg/m^3
$\rho_{layer-2}$	=	Density of layer-2, m/L^3, kg/m^3
$\rho_{layer-3}$	=	Density of layer-2, m/L^3, kg/m^3
$\rho_{reservoir}$	=	Density of layer-2, m/L^3, kg/m^3
v	=	Poisson's ratio
σ_s	=	Initial plastic strength, F/L^2, Pa
ϕ	=	Frictional angle, °
$\dot{\bar{\varepsilon}}^{cr}$	=	Equivalent creep strain rate, t^{-1}, s^{-1}
$\bar{\sigma}^{cr}$	=	von Mises equivalent stress, F/L^2, Pa

REFERENCES

Aadnoy, B.S.: *Modern well design.* Balkema Publications, Rotterdam, The Netherlands, 1996.

Abaqus User's Manual, Vol. 2: Analysis, Version 6.10, 2010. Vélizy-Villacoublay, France: Dassault Systems, 10.2.1-1–10.2.3-10.

Abaqus User's Manual, Vol. 3: Materials, Version 6.8, 2008. Vélizy-Villacoublay, France: Dassault Systems, 19.3.1-17–19.3.2-14.

American Petroleum Institute: Bulletin on formulas and calculations for casing, tubing, drill pipe, and line pipe properties, Bull. 5C3, Fourth Edition, API, Dallas, TX, USA, February 1985.

American Petroleum Institute: Bulletin on performance properties of casing, tubing, and drill pipe, Bull. 5C2, 18th Edition, API, Dallas, TX, USA, March 1982.

Bickley, M.C. and Curry, W.E.: Designing wells for subsidence in the greater Ekofisk area. Paper SPE 24966 presented at the European Petroleum Conference, Cannes, France, 16–18 November 1992.

Bruno, M.S.: Subsidence-induced well failure. *SPEDE* 7:2 (1992), pp. 148–152.

Bruno, M.S.: Geomechanical analysis and decision analysis for mitigating compaction related casing damage. Paper SPE 71695 presented at the 2001 SPE Annual Technical Conference and Exhibition, New Orleans, LA, USA, September 30-October 3, 2001.

Capasso, G. and Mantica, S.: Numerical simulation of compaction and subsidence using ABAQUS. ABAQUS Users' Conference (2006), Boston, MA, USA.

Cipolla, C.L., Hansen, K.K. and Ginty, W.R.: Fracture treatment design and execution in low porosity chalk reservoirs. *SPEPO* 22:1 (2007). pp. 94–106.

Crandall, S.H. and Dahl, N.C.: *An introduction to the mechanics of solids.* McGraw-Hill Book Company, New York, NY, USA, 1959.

Dale, B.D., Narahara, G.M. and Stevens, R.M.: A case history of reservoir subsidence and wellbore damage management in the South Belridge Diatomite field. Presented at the Western Regional Meeting, Anchorage, AK, USA, 22–24, May 1996.

Finol, A. and Farouq Ali, S.M.: Numerical simulation of oil production with simultaneous ground subsidence. *Soc. Petrol. Eng. J.* 15:5 (1975), pp. 411–424.

Gere, J.M. and Timoshenko, S.P.: *Mechanics of materials.* Van Nost. Reinhold, New York, USA, 1987.

Hickman, R.J.: *Formulation and implementation of a constitutive model for soft rock.* PhD Thesis, Virginia Polytechnic Institute and State University, Blacksburg, VA, USA, 2004.

Hilbert, L.B., Gwinn, R.L., Moroney, T.A. and Deitrick, G.L.: Field-scale and wellbore modeling of compaction-induced casing failures. Paper SPE/ISRM 47391 presented at the SPE/ISRM Eurock '98, Trondheim, Norway, 8–10 July 1998.

Ibekwe, I.A., Coker, O.D., Fuh, G.F. and Actis, S.C.: Magnolia casing design for compaction. Paper SPE 79816 presented at the SPE/IADC Drilling Conference, Amsterdam, The Netherlands, 19–21 February 2003.

Kaiser, T.: Post-yield material characterization for strain based design. *SPEJ* 14:1 (2009), pp. 128–134.

Lewis, R.W., Makurat, A. and Pao W.K.S.: Fully coupled modeling of seabed subsidence and reservoir compaction of North Sea oil fields. *Hydrogeology J.* 11:1 (2003), pp. 142–157.

Li, X., Mitchum, F.L., Bruno, M., Padillo, P.D. and Wilson, S.M.: Compaction, subsidence, and associated casing damage and well failure assessment for the Gulf of Mexico Shelf Matagorda Island 623 field. Paper SPE 84553 presented at the SPE Annual Technical Conference and Exposition, Denver, CO, USA, 5–8 October, 2003.

Mitchell, R.F. (ed.): *Petroleum engineering handbook,* volume II *Drilling engineering.* Society of Petroleum Engineers, Richardson, TX, USA, 2006.

Mitchell, R.F. and Goodman, M.A.: Permafrost thaw-subsidence casing design. *JPT* 30:3, 1978.

Sulak, R.M.: Ekofisk Field: The first 20 years. *JPT* 43:10 (1991). pp. 1265–1271.

Sulak, R.M. and Danielsen, J.: Reservoir aspects of Ekofisk subsidence. *JPT* 41:7 (1989): pp. 709–716.

Rabia, H.: *Fundamentals of casing design.* Graham & Trotman, London, 1987.

Rieke, H.H. and Chilingarian, G.V.: *Compaction of argillaceous sediments, developments in sedimentology.* Vol. 16, Elsevier Scientific Publishing Co., New York, USA, 1974.

Roulffignac, E.P., Bondor, P.L., Karanikas, J.M. and Hara, S.K.: Subsidence and well failure in the South Belridge diatomite field. Paper SPE 29626 presented at the Western Regional Meeting, Bakersfield, CA, USA, 8–10 March 1995.

Sayers, C., den Boer, L., Lee, D. Hooyman, P. and Lawrence, R.: Predicting reservoir compaction and casing deformation in deepwater tubidites using a 3D mechanical earth model. Paper SPE 103926 presented at the First International Oil Conference and Exhibition, Cancun, Mexico, August 31-September 2, 2006.

Smith R.E. and Clegg, M.W.: Analysis and design of production wells through thick permafrost. *Proceedings 8th World Pet. Congress,* Moscow. June 1971.

Smith, W.S., Nair, K. and Smith, R.E.: Sample disturbance and thaw consolidation of a deep sand permafrost, *Proc. Permafrost Second International Conference,* Yakutsk, U.S.S.R., 1973.

Weiner, P.D., Wooley, G.R., Coyne, P.L., and Christman, S.A.: Casing strain tests of 13-3/8-in, N-80 buttress connections. Paper SPE 5598 presented at 50th Annual Fall Meeting of SPE-AIME, Dallas, TX, USA, October, 1975.

Yudovich, A., Chin, L.Y. and Morgan, D.R.: Casing deformation in Ekofisk. *JPT* 41:7 (1989), pp. 729–734.

CHAPTER 4

Numerical scheme for calculation of shear failure gradient of wellbore and its applications

Xinpu Shen, Mao Bai & Russell Smith

4.1 INTRODUCTION

The mud weight window (MWW) is the range of mud density values that provides safe support to the wellbore during the drilling process at a given depth. If the mud weight value selected is within the MWW range, the wellbore is stable and no plastic deformation will occur on the wellbore surface. Furthermore, when a safe mud weight is selected within the MWW, no mud loss will occur. The MWW is defined by two bounds. Its lower bound is the shear failure gradient (SFG), which is the minimum mud weight required to avoid plastic failure of the wellbore; its upper bound is the fracture gradient (FG), which is the maximum mud weight value that will not induce fracture openings. Because natural fractures usually exist within various kinds of formations and wellbores are usually vertical, in practice, the value of the minimum horizontal stress is used as the value of FG.

In practice, the MWW of a given wellbore can be designed with either a one-dimensional (1D) analytical method or a three-dimensional (3D) numerical finite-element method (FEM). The 1D method determines horizontal stress components in terms of overburden stress and logging data along the wellbore trajectory, and only the information along the wellbore trajectory is used to determine the MWW (Jones et al., 2007; Charlez 1999). The FE method uses a 3D model that consists of 3D geometry and a 3D mechanical constitutive relationship.

The advantage of 1D analytical tools is that they are highly efficient. Their major disadvantage is that several assumptions are adopted with the input data. These assumptions are usually reasonable, but they may not be accurate enough for some cases with complicated geological conditions, such as subsalt wells. In the past 20 years, various studies focused on increasing the accuracy of 1D solutions (Keaney et al., 2010; Doyle et al., 2003; Constant and Bourgoyne 1988; and Abousleiman et al., 2009). The advantage of the 3D numerical method is that it can accurately calculate the geostress distribution within formations by a 3D FE method. Its major disadvantage is that it is not as efficient as the 1D method. The 3D method has been increasingly applied in recent years (Islam et al., 2009; Shen 2009; Shen et al., 2010; and Shen et al., 2011).

The FG calculation is rather straightforward; it can be performed by using the minimum horizontal stress value at the wellbore location. The SFG calculation, however, is a rather complicated process. It requires the calculation of the stress distribution around the wellbore, and the determination of the stress state in terms of plastic loading criteria and values of stress components, such as that described by Zoback (2007).

In the following sections, the numerical scheme for the calculation of SFG is first presented. A benchmark solution for a given wellbore within isotropic formations is presented afterward, and a numerical solution of SFG and its comparison with results obtained by Drillworks software are presented. At the end of the chapter, an example of the application of the proposed scheme is used to calculate the SFG for the well section at the interface between the salt and reservoir at the saltbase formation.

4.2   SCHEME FOR CALCULATION OF SFG WITH 3D FEM

As shown in Fig. 4.1, the mud weight pressure will be applied to the inner surface of the borehole. The width, height, and thickness of the submodel used here are w = 6 m, H = 9 m, and T = 6 m, respectively. The boundary conditions were obtained from the numerical results of the global model.

In the calculation of the mud weight pressure, a reference pressure P_w, which is generated by static water, was calculated first. Next, a loading factor α (α>1) was assumed. This loading factor actually represents the mud weight pressure, and its value is the specific weight of the mud. Here, α is set at $\alpha = 2$. With a hydraulic pressure of $P = 2P_w$, the wellbore is stable and no plastic deformation occurs.

The calculation of the minimum safe mud weight pressure determines the minimum loading factor β with which the hydraulic pressure can maintain wellbore stability. Below this minimum value, plastic deformation will occur to the wellbore surface. To calculate the minimum safe mud weight, a negative hydraulic pressure with loading δ was added incrementally to the reference pressure αP_w. Therefore,

$$\delta = \sum_{i=1}^{N} \Delta\delta_i, \ \beta = \alpha - \delta \tag{4.1}$$

where δ_i was set at <5% in this calculation, and N will be calculated automatically in the incremental calculation process with FEM. When plastic deformation occurs, the calculation stops because a given criterion was satisfied. The resultant minimum mud weight pressure required for wellbore stability will be:

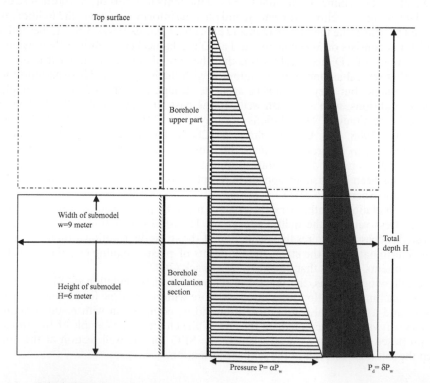

Figure 4.1.   Calculation scheme for SFG.

$$P = \beta P_w \qquad (4.2)$$

In this calculation, the pressure gradient over the submodel section is assumed to be constant. Consequently, the resultant mud weight gradient is a section-averaged value.

The P_w value is calculated with reference to the total depth of a given section; the resultant mud weight pressure gradient is the conventional equivalent mud weight gradient. The mud weight pressure value at a given point can be obtained by multiplying the depth at that point by the related pressure gradient value.

4.3   NUMERICAL SOLUTION OF SFG AND ITS COMPARISON WITH RESULTS OBTAINED BY DRILLWORKS

As described in the previous section, the SFG is the minimum value of a mud weight pressure gradient for a safe wellbore that precludes mechanical collapse in formations. Drillworks™ calculates the SFG on the basis of the following:

- Overburden gradient (OBG), which is the sum of the effective vertical stress gradient (σ_z) and pore pressure gradient (pp)
- Fracture gradient (FG), which is the minimum horizontal stress component (ShG)
- Maximum horizontal stress component (SHG)
- Strength parameters, which are cohesive strength c and internal friction angle ϕ if the Mohr-Coulomb criterion is adopted. Because the analytical stress solution of a plane strain cylinder under a given geostress and inner pressure was used to derive the SFG, the SFG solution obtained by Drillworks is basically a 1D solution.

In an Abaqus finite element calculation, SFG can be calculated in a similar, but not identical, manner. Usually, for a drilling related simulation, the FEM calculations must accomplish the following:

- Construct an initial geostress field and an initial pore pressure field.
- Remove the drilled elements and apply a mud weight pressure on the inner surface of the wellbore.
- Perform elastoplastic deformation calculations coupled with porous fluid flow.

The proposed pore pressure field is given as a static one, i.e., no time-related variation occurs in the pore pressure field, and the value of the hydrostatic pore pressure is assumed to be the one as produced by static water for a given depth. Consequently, the pore pressure gradient is:

$$p_w = 1 \, g/cc \qquad (4.3)$$

The elastoplastic calculation is performed in the effective stress space.

In the following example, the Abaqus calculation of SFG is introduced first; the result of the SFG calculation from Drillworks will be introduced and a comparison between the two will be made.

Because the focus of the analysis is placed on the wellbore stability in the drilling process, no pore pressure depletion is considered. The mud weight pressure during the calculation is obtained from drilling data contained in reference documents.

4.3.1   *The model geometry of the benchmark and its FEM mesh*

Figure 4.2 illustrates the geometry of the benchmark at the field scale. For the model block, width w = 800 m, height H = 1100 m, and thickness T = 800 m. The wellbore trajectory contains a 500 m vertical section, which is connected to a ¼ circular section with radius R = 500 m (at approximately 5°/100 ft). The lower end is a horizontal wellbore section of 100 m. For simplicity, all formations are assumed to be elastoplastic porous media. In practice, this model can accommodate faults and other structural elements.

Fig. 4.3 shows the outline of the geometry of the global model, which is a block containing the entire wellbore. In comparison with the borehole size, the global model is too large to make an analysis with reasonable precision. To overcome this difficulty, this calculation uses the submodeling technique. By using a submodeling step, a model with finer mesh around the borehole can be built with which the boundary conditions can be obtained from the results of the global model.

Details about submodeling can be found in Chapter 3 and in the Abaqus Analysis User's Manual (Dassault Systems 2008).

Fig. 4.4 shows a submodel. The shape of borehole has been added to the block, and a finer mesh was adopted for the area around the borehole. The loads used by the submodel are the same as those of the global model. Node-based submodeling technique is adopted in the process of transferring displacement results of the global model to the boundary of the submodel.

The geometry of the model can be created in terms of the shape of the real wellbore section. The coordinate of the central point of the submodel can be determined by aligning the submodel to the target position within the global model. In the process, the submodel can

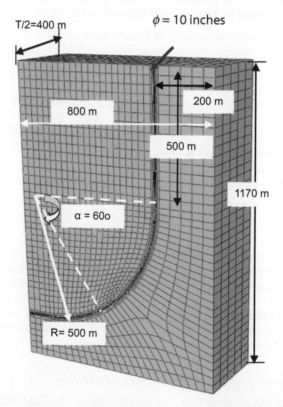

Figure 4.2. Well path and point at deviated section of the benchmark: a block with a curved well trajectory.

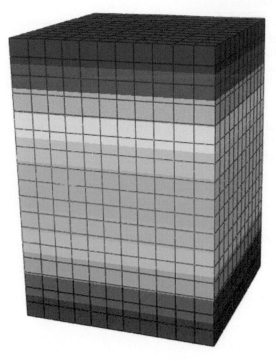

Figure 4.3.   The mesh of global model used in the calculation.

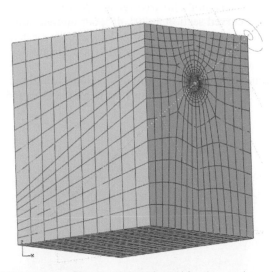

Figure 4.4.   Submodel illustration: a finer mesh was adopted for the area immediately around the bore-hole (submodel for deviated well section at $\alpha = 60°$).

be generated by eliminating all other parts from the global model and adding the details of the geometry. Alternatively, a submodel can be created independently of the global model at first, and then positioned at the target place in the global model. Both methods can create a good submodel.

4.3.2   *Loads and parameters of material properties*

For the global model, a water-derived hydraulic pressure and a gravity field are the major loads applied. A tectonic geostress field was constructed with a gravity load and the following lateral tectonic factors:

$$S_H = 0.9\sigma_v, \quad S_h = 0.85\sigma_v \tag{4.4}$$

where σ_v is the vertical stress produced by gravity with a density value as $\rho = 2.1\ g/cc$.

On the inner surface of the borehole, a reference hydraulic load corresponding to water pressure has been applied.

The parameter values of the material properties are assumed as follows:

```
**
** MATERIALS
**
*Elastic Young's modulus and Poisson's ratio
 1e+09 Pa, 0.25
*Mohr Coulomb internal friction angle and dilatancy angle
 25., 22.5
*Mohr  Coulomb  Hardening:  strength  variation  with  plastic
strain:
 0.5e+06 Pa, 0.
 0.49998e+06 Pa, 0.5
**
```

4.3.3   *Abaqus submodel calculation and results with Mohr-Coulomb model*

The submodels used in the calculation for true vertical depth (TVD) = 300 m, 500 m at vertical section, and points of curved section at 30°, 45°, 60° and 90° angle points are shown in Fig. 4.5 to Fig. 4.11 respectively. Fig. 4.5 and Fig. 4.6 show the mesh and loading conditions of the submodel at TVD = 300 m. Fig. 4.7 to Fig. 4.11 show the meshes of submodels at other positions mentioned above.

The resultant equivalent mud weight pressure gradient obtained by Abaqus for submodels at depth 27 m, 300 m, 500 m, 750 m (30° angle), 853 m (45°), 933 m (60°), and

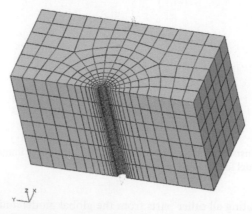

Figure 4.5.   Submodel adopted for the vertical section at TVD = 300 m: mesh, c3d20r element (i.e., second order 3D element with 20 nodes and reduced integration scheme) was used to "discretize" the model. Only half of the model is shown here because of symmetry.

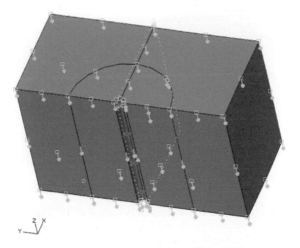

Figure 4.6. Submodel adopted for the vertical section at TVD = 300 m: loads and boundary conditions. Only half of the model is shown here because of symmetry.

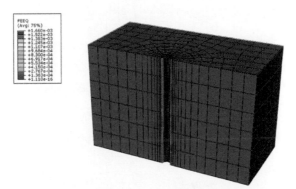

Figure 4.7. Submodel adopted for the vertical section at TVD = 300 m: plastic strain initiation. Only half of the model is shown here because of symmetry.

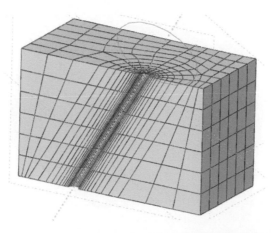

Figure 4.8. Submodel adopted for the curved trajectory section of 30° angles. Only half of the model is shown here because of symmetry.

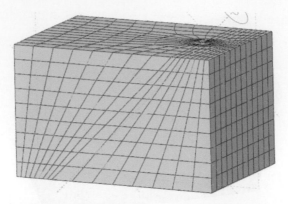

Figure 4.9.   Submodel adopted for the curved trajectory section of 45° angles.

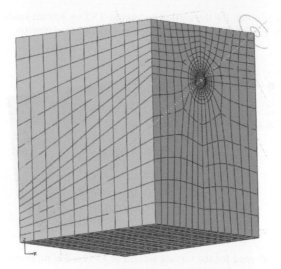

Figure 4.10.   Submodel adopted for curved trajectory section of the 60° angles.

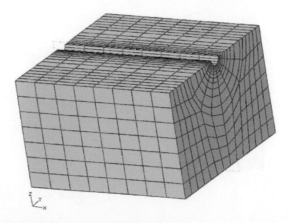

Figure 4.11.   Submodel adopted for the curved trajectory section of the 90° angles. Only half of the model is shown here because of symmetry.

Table 4.1. Comparison of Abaqus numerical results with Drillworks results.

TVD/m	MW/g/cc: Abaqus results	MW/g/cc: Drillworks results	Remarks
0	0	0	
27	0	0	Vertical section
300	1.4088	1.39	Vertical section
500	1.4588	1.452	Vertical section
750	1.5088	1.55	30°
853.5	1.658	1.605	45°
933.	1.6177	1.65	60°
1000	1.6088	1.678	90°

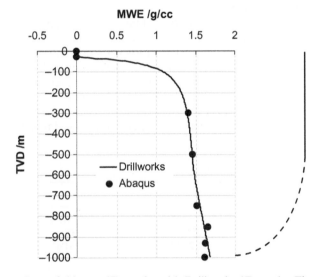

Figure 4.12. Comparison of Abaqus 3D results with Drillworks 1D results. The curve on the right shows the well path trajectory.

1000 m (90° angle, which is a horizontal well section) are listed in Table 4.1 and illustrated in Fig. 4.12.

From Fig. 4.12, it is seen that the minimum safe equivalent mud weight gradients, which is the SFG obtained numerically with Abaqus, are the same as those obtained by Drillworks for points in vertical borehole sections. They are very near to one another for those points along the deviated well path. The Abaqus result at the 45° position is approximately 5% greater than the Drillworks solution, and the result at the 90° section is approximately 5% less than the Drillworks solution.

4.3.4 *Results comparison with Drucker-Prager criterion between Abaqus and Drillworks*

To validate the accuracy of results obtained with Drillworks and those obtained with Abaqus with additional plastic yielding criteria, the Drucker-Prager yielding criterion was used in the following calculation of the minimum safe mud weight.

According to Dassault Systems (2008), to match the Drucker-Prager criterion with the Mohr-Coulomb criterion on the testing results obtained with both uniaxial compression and uniaxial tension cases, the following mathematical relationship given in Eq. 4.5 and Eq. 4.6 exists between the strength parameters of the Mohr-Coulomb and Drucker-Prager criteria.

By comparing the Mohr-Coulomb model with the linear Drucker-Prager model described in Dassault Systems (2008), the following relationship can be obtained:

$$\tan\beta = \frac{6\sin\phi}{3-\sin\phi}, \quad \sigma_c^0 = 2c\frac{\cos\phi}{1-\sin\phi} \tag{4.5}$$

$$K = \frac{3-\sin\phi}{3+\sin\phi} \tag{4.6}$$

These expressions for β, K, and σ_c^0 define the match between the values of the parameters of the two models in triaxial compression and tension.

From Eq. 4.5 and Eq. 4.6, the values of the parameters for both the Mohr-Coulomb and Drucker-Prager criteria are listed in Table 4.2.

Fig. 4.13 shows the numerical results of SFG obtained with Abaqus. The round dot represents the numerical results obtained by using the Mohr-Coulomb yielding criterion. The square mark represents the result obtained by using the Drucker-Prager model. Fig. 4.13 shows that the numerical results of the SFG values are near but slightly larger than those obtained with Drillworks at most points along the well path; however, the numerical results are obviously larger than the Drillworks solution for the point at the 60° curve.

The results which represented by a thin-red line with triangle marks shown in Fig. 4.13 are obtained with jointed material model. The Drucker-Prager yielding criterion was used in

Table 4.2. Values of parameters.

	M-C	D-P
c	0.5 MPa	1.57 MPa
/phi	25°	44.53°
C2	1.25 MPa	4.0 MPa

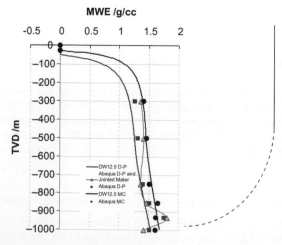

Figure 4.13. Comparison of Abaqus 3D results with Drillworks 1D results.

this set of results. In the jointed material model, a joint in the horizontal surface is assumed. The result for the point at the 60° curve is near the previous value generated by the Drucker-Prager solution at the same position, but the value is slightly larger than the one set of numerical results.

4.3.5 *Remarks*

In this calculation, the Abaqus finite element software was used to calculate the safe lower bound of mud weight pressure. The results obtained with Abaqus were displayed and compared with the solution obtained from the 1D Drillworks software. The Abaqus results are very near the Drillworks solutions at most sections of a given curved wellbore.

Some insignificant differences between the two sets of solutions were observed for points along the deviated portion of the well path.

4.4 COMPARISON OF ACCURACY OF STRESS SOLUTION OF A CYLINDER OBTAINED BY ABAQUS AND ITS ANALYTICAL SOLUTION

The following figures show the model of a plane strain cylinder under both internal and external pressures.

Elastic behavior is assumed by the model. The internal pressure is represented by $p_i = 100$ Pa, and the external pressure is represented by $p_i = 200$ Pa.

Fig. 4.16 shows that the numerical solution obtained with first order elements (c2d4r) has an approximate 10% deviation from the corresponding analytical solution, and the results obtained with second order elements (c2d8r) has a deviation from the analytical solution of only approximately 1%.

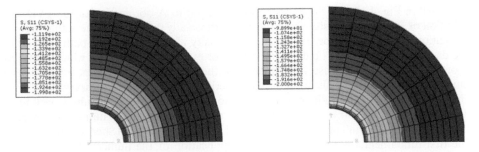

Figure 4.14.   Comparison of radial stress component σ_r obtained by Abaqus with first order element (left) and second order elements (right).

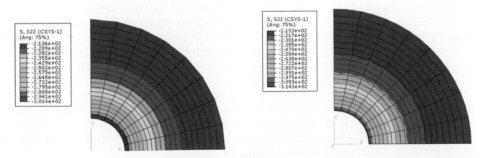

Figure 4.15.   Comparison of circular stress components σ_θ obtained by Abaqus with first order element (left) and second order elements (right).

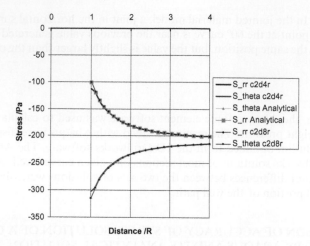

Figure 4.16.   Comparison between numerical solutions and analytical solutions (Cauchy solutions).

This result indicates that it is necessary to use second order elements in numerical analyses and that the numerical method can obtain solutions with a rather high degree of accuracy.

4.5   APPLICATION

Subsalt wells are common in the northern region of the Gulf of Mexico (Shen 2009; Shen *et al.,* 2010).

As shown in Fig. 4.17, the overburden gradient at point A obtained with the TVD axial distribution of formation density along the well path usually differs from the actual overburden situation, which should be calculated with density along the vertical path represented by the dashed line. This is particularly important for extended-reach wells, which cross long horizontal distances. In this case, a 1D analysis of the wellbore stability appears to be insufficiently accurate, and the application of 3D numerical tools becomes necessary.

The variation of water depth over a certain horizontal distance of a subsalt well in the northern region of the Gulf of Mexico is another reason for using a 3D numerical tool along with conventional 1D analytical tools, including Drillworks software (Shen 2009).

The first fundamental issue for a successful application of the numerical method in the subsalt wellbore stability analysis is the accurate input of the pore pressure encountered along the wellbore. A second issue affecting its success is the accurate description of the 3D structure of the salt body and related formations, which will be involved in the calculation.

Drillworks software is a set of 1D tools for analysis used for pore pressure prediction and wellbore stability analysis. It is referred to as a 1D tool because it analyzes wellbore stability and pore pressure based on TVD information along the well path. To be more precise, Drillworks software is based on the linear elastic theory and the 2D plane strain conditions.

In practice, Drillworks software has proven to be a very powerful analytical tool in the prediction of pore pressure for many wellbores in the Gulf of Mexico and in other locations, and has laid a good foundation for a successful 3D numerical analysis of subsalt wellbore stability. Conversely, modern seismic technology can provide accurate spatial geological and structural information for the salt dome/canopy and related formations, as shown in Fig. 4.17. Consequently, it is not only necessary, but also practical to perform a 3D numerical analysis in relation to subsalt wellbore stability.

The primary task of this chapter is to perform a pore pressure prediction with Drillworks software for the Wellbore A-29 at Viosca Knoll. A 3D numerical analysis with FEM software

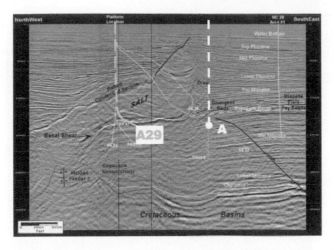

Figure 4.17.   Seismic image of salt structure, formation, and relative position of Wellbore A-29 (Holly *et al.*, 2004).

will be presented for a given depth position at the salt exit of the wellbore. This method for sub-salt wellbore stability analysis combines conventional 1D analytical tools, such as Drillworks software (Shen 2009), with a 3D numerical tool, such as Abaqus (Dassault Systems 2008).

4.5.1   *Pore pressure analysis with Drillworks*

Wellbore A-29 is located in the Viosca Knoll field and the pay zones are directly below the Pompano salt. The pore pressure gradients were estimated and were calibrated to all of the available data, such as mud weight, equivalent circulating density (ECD), and well events. Fig. 4.18 shows the results from deriving pore pressure and local stress regime at the VK989 Well A-29 by means of analyzing wireline logs.

In the first track on the left of Fig. 4.18 is a GR curve with a shale base line (SHBL) that was used to distinguish the shale from other lithological formation. The second track shows the resistivity data, the shale picks derived from the GR curve in track 1, filtered resistivity data using the selected shale points, and a normal compaction trend line (NCTL) that was used in an Eaton resistivity analysis to generate the pore pressure curve (green) displayed in the right first track.

The sonic analysis is shown in the third track with the NCTL coming from the Bowers method; the curves for a density analysis are in the fourth track.

The right track of the display in Fig. 4.18 shows the curves relating to the pressure analysis and the drilling of the well, the casing program, overburden gradient (OBG), fracture gradient (using a Matthews and Kelly relationship with $K_0 = 0.8$), and mud weight (black line). Three pore pressure curves are included on this track. Pore pressure (dark green) is derived from the resistivity, and the lighter green curve is derived from the sonic data. The value of the pore pressure gradient at the salt exit point of Well A-29 was calculated at 17,497 Pa/m (14.9 ppg).

Because the curves derived from the various petrophysical tools and the equations that derive a pore pressure from them do not agree, the resulting challenge is to determine the true pore pressure at any given point in the well. This interpreted curve is shown on the plot as a light green line with square marks. This curve demonstrates the "best fit" for all of the data, considering the different properties that are measured by the various tools, the calibration to the measured pressures in the sands, and the well events that occurred while drilling (not shown). In the deep part of the well, the sand pressures and the Rhob (represents density) and DT (represents sonic data) pressures were monitored more closely. In the shallow

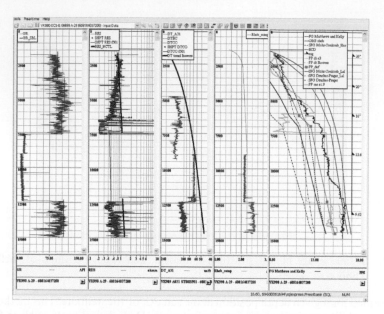

Figure 4.18.   Wellbore A-29 in VK989 field area: pore pressure and stress analyses.

part of the well, the resistivity and sonic data-derived pressures were considered to be more valid because the DT-derived pressures are above the mud weight that was used to drill the well. This interpreted pore pressure was used in subsequent modeling of the wellbore stability conditions.

4.5.2   The 3D computational model

The stress distribution around the salt body is directly related to both the geometry and relative position of the entire salt body. To obtain accurate stress field information around the salt, a mechanical analysis at the field scale must be performed. Only then is it possible to further present the correct boundary conditions for the wellbore stability analysis of subsalt wells.

Abaqus submodeling techniques were used to manage the field-to-reservoir scale discrepancy. The concept of the submodeling technique includes using a large scale global model to produce boundary conditions for a smaller scale submodel.

Several references (Harrison *et al.,* 2004; Ertekin and Karpyn 2005; Marinez *et al.,* 2008; Aburto and Clyde 2009) show that the point of exit from a salt body is the weakest part of a subsalt wellbore. This weakness occurs because the stiffness of salt compared to that of normal sedimentary rocks is very different; consequently, it cause a jump or displacement force across the interface of these two materials. This scenario can also result in unfavorable mechanical loads to the wellbore and on the casing, if any.

In this calculation, a global model at the field scale was established first to simulate the structural influence of salt geometry (see Fig. 4.19). A submodel was built for the wellbore section where it exits the lower surface of the salt body at its center (see Fig. 4.21 for the position). Wellbore sections in both salt and subsalt formations are included in this submodel (Fig. 4.22). The TVD depth (i.e., the distance from center of the submodel to a vertical point at the sea floor) of the submodel is 3595 m.

4.5.2.1   *Global model: Geometry, boundary condition, and loads*
Fig. 4.19 shows the geometric profile of the model. Its width and thickness are both 10 km; its height on the left side is 10 km, and its height on the right is 9.6 km, which shows a variation in water depth.

According to information presented in the references (Harrison *et al.*, 2004; Ertekin and Karpyn 2005), the geometry of the salt body was built as shown in Fig. 4.20. The outer edge of the pancake-shaped salt body has the diameter of 7.01 km, and the maximum thickness is 1.676 km. Its upper surface has a 30° angle with the horizon. The depth of its top edge from the sea floor is 1.219 km. The geometry and relative position of the salt body is shown in Fig. 4.21.

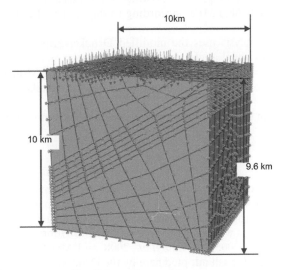

Figure 4.19.   Model geometry: profile of the entire model.

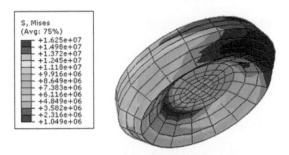

Figure 4.20.   Model geometry: shape of the salt body.

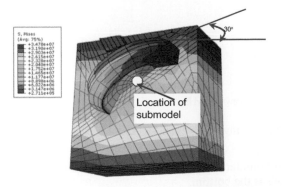

Figure 4.21.   Model geometry: location of the salt body in the global model.

As shown in Fig. 4.19, the loads sustained by the model include the following:

- Hydrostatic pressure produced by sea water at the top of the model
- Gravity load distributed within the model body
- Pore pressure distributed in the formations

Because the salt body has no porosity or permeability, the pore pressure within the salt body is assumed to be zero. Pore pressure existing in the subsalt reservoir formation is given as 9,950 psi (approximately 68.6 MPa), according to the predicted pore pressure value from Drillworks software.

For a reservoir, this analysis uses the modified Drucker-Prager yielding criterion. The cohesive strength and frictional angle of the Drucker-Prager model are given the following values: $d = 1.56$ Mpa, $\beta = 44°$, which correspond to values in the Mohr-Coulomb model as $c = 0.5$ Mpa, $\phi = 25°$ (For additional information about the details of the two models, see Shen 2009.)

The creep law, given in the following equation (Dassault Systems 2008), is adopted by:

$$\dot{\bar{\varepsilon}}^{cr} = A\left(\bar{\sigma}^{cr}\right)^{n} t^{m} \tag{4.7}$$

where $\dot{\bar{\varepsilon}}^{cr}$ represents the equivalent creep strain rate; $\bar{\sigma}^{cr}$ represents the von Mises equivalent stress, and t is a total time variable. A, n, and m are three model parameters.

The values of material properties are provided in the following list; references are made to the data listed in Table 4.2 as well as to the data reported in Hunter *et al.*, 2009, Infante and Chenevert 1989, Maia *et al.*, 2005, Marinez *et al.*, 2008, and Aburto and Clyde 2009. The values of strength parameters for salt adopted here by the Drucker-Prager model are $d = 4$ Mpa, $\beta = 44°$, which correspond to values in the Mohr-Coulomb model as $c = 1.25$ Mpa, $\phi = 25°$.

List of values of the material parameters is given as follows:

*Material, name=reservoir
*Density in kg/m³
2300.,
*Elastic modulus in Pa
1e+10, 0.3
*Drucker Prager
44., 1., 30.
*Drucker Prager Creep, law=TIME
2.5e-22, 2.942, -0.2
*Drucker Prager Hardening, type=SHEAR, in Pa
1.56e+06,0.
*Material, name=salt
*Density in kg/m³
2100.,
*Elastic modulus in Pa
1e+10, 0.3
*Drucker Prager
44., 1., 30.
*Drucker Prager Creep, law=TIME
2.5e-22, 2.942, -0.2
*Drucker Prager Hardening, type=SHEAR, in Pa
4.e+06,0.

Zero displacement boundary constraints were applied in the normal direction of the four lateral surfaces as well as the bottom.

To simplify the model, the details of the geometry of the reservoir formation are modeled on its top surface, i.e., the shape of the interface of between the salt body and formation.

Other parts of the geometrical characteristics of the reservoir formation have been omitted in the model. Because the dynamic effect of porous flow is not considered in the calculation, this simplification will have no negative influence on the values of mud weight window.

4.5.2.2 *Numerical results of the global model*

Fig. 4.22 to Fig. 4.24 show the numerical results of the global model, which are used as a base for the wellbore stability analysis performed with the submodel. The distribution of a vertical stress component, shown in Fig. 4.8, has a reasonable accuracy with reference to the hydraulic load on the top of the model and the gravity load within the model body.

Fig. 4.23 shows the initial field of geostrain. The value of geostrain is designed to be as small as possible to minimize the original strain and to focus on the strain relevant to drilling activities.

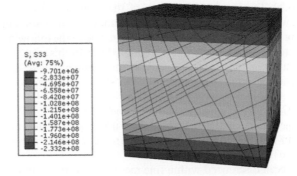

Figure 4.22. Numerical results of the global model: distribution of vertical stress.

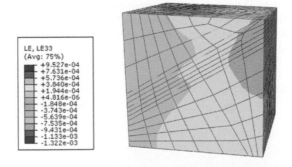

Figure 4.23. Numerical results of global model: distribution of vertical strain.

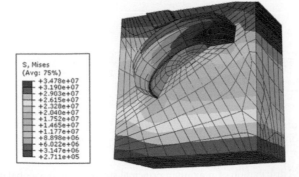

Figure 4.24. Numerical results of the global model: distribution of deviatoric stress.

Fig. 4.24 shows the distribution of deviatoric stress (i.e., von Mises stress); the differences between the three principal stress components are small within the salt body, which has rather high creep fluidity.

Fig. 4.22 to Fig. 4.24 indicate that the numerical results of the global model are reasonable and, consequently, can be used in the submodeling calculation.

4.5.2.3   *Vector-distribution of principal stresses*

Fig. 4.25 shows the vector distribution of the principal stresses. The optimized trajectory is arranged to be perpendicular to the vector of maximum principal stress and the minimum principal stress. Wellbore stability is proved to be the most favored among all possible choices (Shen 2009). The adoption of this optimized trajectory will also present the stiffness symmetry to the wellbore at the salt exit, although the geostress field has no symmetrical property at the same place. Fig. 4.25 also shows the vectoric distribution of the maximum principal stress within the global model. To provide a clear view of the vector distribution, the Y-Z sectional view at the well path was selected for display. The sign convention of solid mechanics has been followed in describing the order of principal stresses here and in the following contexts in which algebraic values of stress components will be accounted for. In this way, the maximum principle stress is the minimum compressive stress, and the minimum principal stress is the maximum compressive stress. The distribution of other principal stresses was neglected for brevity.

4.5.2.4   *Submodel: Geometry, boundary condition, and loads*

Fig. 4.26 shows the geometry of the submodel. The upper part is salt, and the lower part is reservoir formation in which a non-zero pore pressure exists. Here, the wellbore axis has an inclination angle of 30° to the vertical.

The submodel has been discretized with a fine mesh around the wellbore: the element edge size in the radius direction is only 1/10 of the wellbore radius (12 in. diameter), as shown in Fig. 4.27.

The loads sustained by the submodel are similar to those sustained by the global model, except the hydrostatic pressure applied to the wellbore surface, which is the mud weight pressure. The numerical scheme for the calculation of a safe mud weight gradient is described in the last section and is omitted here.

Figure 4.25.   Vectoral distribution of maximum principal stress within the global model.

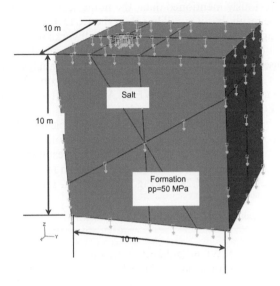

Figure 4.26.   Model geometry: profile of the submodel.

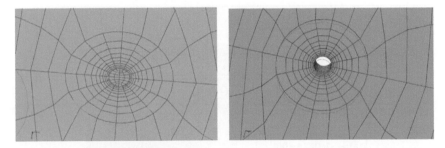

Figure 4.27.   Submodel geometry: mesh around the borehole (a) initial mesh, and (b) borehole was drilled.

The boundary conditions of the submodel are taken from the numerical results of the global model. Displacement constraints are applied on all lateral surfaces. Because of the displacement in the radius direction around the borehole, top and bottom displacement constraints were applied in the normal direction to these two surfaces; the values of these displacements were also taken from the global model, which is usually not zero.

4.5.2.5   *Numerical results of the submodel*
The numerical model of the SFG safe mud weight pressure obtained with the submodel is provided in Eq. 4.8:

$$p_{mw} = 59.292 \, \text{MPa} \qquad (4.8)$$

Transfer the above pressure further into equivalent pressure gradient in ppg as

$$p_{mwe} = 17794/1174.295 = 15.15 \, \text{ppg} \qquad (4.9)$$

The results of the SFG, which is the lower bound of the MWE obtained with Drill-works software, are listed in Table 4.3. Table 4.3 shows the values of various MWE values.

Compared with the previously mentioned data, the numerical result obtained with Abaqus is the most conservative result, but it is still less than the ECD. It can also be concluded from Table 4.3 that the strength correlation model used in Drillworks software from sonic data is too risky; it should not be used in subsalt basins because the sonic data from salt is normally not accurate and, consequently, should not be used directly in analysis. The adjusted MWE result from Drillworks software is far better than the original one. With the same strength parameters values, the numerical result of MWE by Abaqus software provides a better value, which is closer to the ECD value than other results.

Conversely, with reference to numerical values listed in Table 4.3, the MWE value of the ECD log is often too conservative and, thus, is not the best choice.

In summary, for the best practice, the adjusted MWE/SFG result from Drillworks software is a reasonable choice. If available, the 3D numerical solution for a typical subsalt wellbore stability will be the best choice. Fig. 4.28 and Fig. 4.31 show the numerical stress distributions

Table 4.3.   Comparison of values of MWE (in Pa/m).

MWE by Drillworks software with CS obtained by Horsrud correlation model	MWE by Drillworks software with manually adjusted CS input	MWE by Abaqus 3D numerical method	MWE by ECD log
15853.0	17379.6	17794.0	19011.9

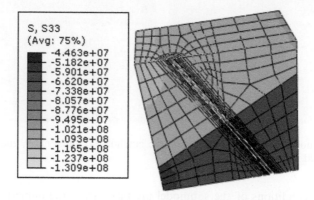

Figure 4.28.   Numerical results of submodel: distribution of vertical stress component.

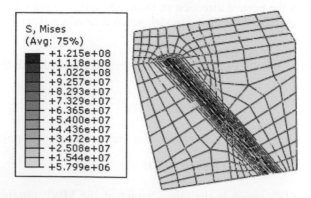

Figure 4.29.   Numerical results of submodel: distribution of deviatoric stress (von Mises stress).

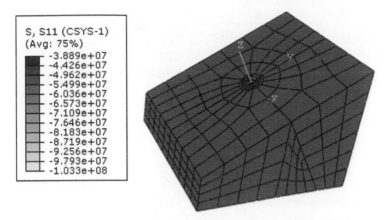

Figure 4.30.   Numerical results of the subsalt part of the submodel: distribution of radius stress around borehole with axis of borehole as its local z-direction; see local coordinates within the figure.

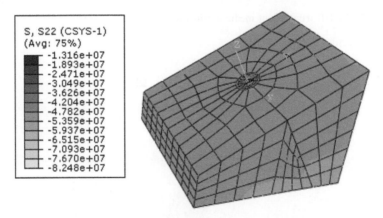

Figure 4.31.   Numerical results of the subsalt part of the submodel: distribution of hoop stress around borehole.

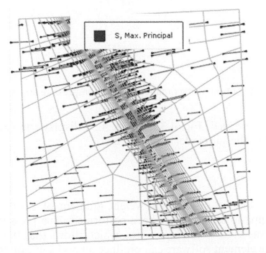

Figure 4.32.   Vectoral distribution of minimum principal compressive stress.

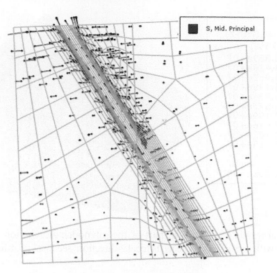

Figure 4.33.   Vectoral distribution of medium principal stress.

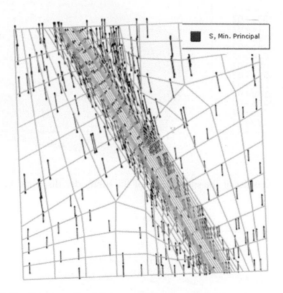

Figure 4.34.   Vectoral distribution of maximum principal compressive stress.

within the submodel described. Fig. 4.32 to Fig. 4.34 display the vectoral distribution of principal stresses within the submodel at the salt exit.

4.6 REMARKS

The 3D numerical calculation of the mud weight gradient of a subsalt Wellbore A-29 at the Viosca Knoll Pompano salt field has been performed. The pressure predicted by using Drillworks software was used as the input data, which shows that the integrated use of 1D software and 3D finite element software can produce a good solution for mud weight design for subsalt wells.

Conversely, in comparison with the results obtained from 1D Drillworks software, the 3D numerical results of a minimum safe mud weight gradient are more conservative, but in practice, the ECD log recorded in practice is too conservative to be an economical solution.

A submodeling technique was used in this calculation that made it possible to perform a 3D numerical analysis that considers both the field scale structural factors and the local environmental factors at the wellbore scale.

For the best practice of determining wellbore stability, adjusted MWE results by Drillworks software is a reasonable choice. If available, a 3D numerical solution for a subsalt WBS would be the most appropriate option.

Because of strict requirements to accurately determine the mud weight gradient to maintain the stability of wellbore section upon exiting a salt body with creep properties, it is often necessary to set casing above this point. The estimate of the integrity of the casing at this specific depth will be the focus of further work.

NOMENCLATURE

A	=	Model parameter in creep model
n	=	Model parameter in creep model
m	=	Model parameter in creep model
α	=	Loading factor
P_w	=	Reference pressure generated by static water, Pa
P	=	Hydraulic pressure, Pa
β	=	Minimum safe loading factor
δ_i	=	Incremental loading factor
σ_z	=	Effective vertical stress gradient, Pa
pp	=	Pore pressure gradient, ppg
FG	=	Fracture gradient, ppg
OBG	=	Overburden gradient, ppg
ShG	=	Minimum horizontal stress component, Pa
c	=	Cohesive strength, Pa
ϕ	=	Internal friction angle, °
FEM	=	Finite Element method
ρ	=	Density, kg/m³
TVD	=	True vertical depth, m
MW	=	Mud weight, g/cc
MWE	=	Mud weight equivalent, g/cc
MWW	=	Mud weight window
σ_c^0	=	Compressive strength used in Drucker-Prager criterion
K	=	A ratio parameter used in Drucker-Prager criterion
d	=	Cohesive strength used in Drucker-Prager criterion, Pa
$\bar{\sigma}^{cr}$	=	von Mises equivalent stress, Pa
$\dot{\bar{\varepsilon}}^{cr}$	=	Eequivalent creep strain rate
p_{mw}	=	Safe mud weight pressure, Pa
SFG	=	Shear fracture gradient
ECD	=	Equivalent circulating density
SHBL	=	Shale base line
NCTL	=	Normal compaction trend line

REFERENCES

Abousleiman, Y., Hoang, S., Ortega, J.A. and Ulm, F.J.: Geomechanics field characterization of the two prolific U.S. Mid-West gas plays with advanced wire-line logging tools. Paper SPE 124428-MS presented at the SPE Annual Technical Conference and Exhibition, New Orleans, LA, USA, 4–7 October, 2009.

Aburto, M. and Clyde, R.: The evolution of rotary steerable practices to drill faster, safer and cheaper deepwater salt sections in the Gulf of Mexico. Paper SPE/IADC 118870 presented at the SPE/IADC Drilling Conference and Exhibition, Amsterdam, The Netherlands, 17–19 March, 2009.

Constant, D.W. and Bourgoyne Jr., A.T.: Fracture-gradient prediction for offshore wells. *SPE Drilling Engineering* 3:2 (1988), pp. 136–140.

Charlez, Ph. A.: The concept of mud weight window applied to complex drilling. Paper SPE 56758 presented at the SPE Annual Technical Conference and Exhibition, Houston, TX, USA, 3–6 October, 1999.

Dassault Systems: 2008. *Abaqus Analysis User's Manual*, USA.

Doyle, E.F., Berry, J.R. and McCormack, N.J.: Plan for surprises: pore pressure challenges during the drilling of a deepwater exploration well in mid-winter in Norway. Paper SPE 79848 presented at the SPE/IADC Drilling Conference, Amsterdam, The Netherlands, 19–21 February, 2003.

Ertekin, T. and Karpyn, Z.T.: Tahoe field case study—understanding reservoir compartmentalization in a channel-levee system. *Gulf Coast Association of Geological Societies Transactions*, Vol. 55. (2005).

Harrison, H., Kuhmichel, L., Heppard, P., Milkov, A.V., Turner, J.C. and Greeley, D.: Base of salt structure and stratigraphy Data and models from Pompano Field, VK 989/990, Gulf of Mexico. *GCSSEPM Foundation 24th Annual Research Conference "Salt-Sediment Interactions and Hydrocarbon Prospectivity: Concepts, Applications, and Case Studies for the 21st Century"* (Edited by P.J. Post, D.L. Olson, K.T. Lyons, S.P. Palmes, P.F. Harrison, and N.C. Rosen), 2004, pp. 141–156.

Hunter, B., Tahmourpour, F. and Faul, R.: Cementing casing strings across salt zones: an overview of global best practices. Paper SPE 122079 presented at the SPE Asia Pacific Oil and Gas Conference & Exhibition, Jakarta, Indonesia, 4–6 August, 2009.

Infante, E.F. and Chenevert, M.E.: Stability of boreholes drilled through salt formations displaying plastic behaviour. *SPE Drilling Engineering* 4:1 (1989), pp. 57–65.

Islam, M.A., Skalle, P., Faruk, A.B.M.O. and Pierre, B.: Analytical and numerical study of consolidation effect on time delayed borehole stability during underbalanced drilling in shale. Paper SPE 127554 presented at the Kuwait International Petroleum Conference and Exhibition, Kuwait City, Kuwait, 14–16 December, 2009.

Jones, J., Matthews, M.D. and Standifird, W.: Novel approach for estimating pore fluid pressures ahead of the drill bit. Paper SPE 104606 presented at the SPE/IADC Drilling Conference, Amsterdam, The Netherlands, 20–22 February, 2007.

Keaney, G., Li, G. and Williams, K.: Improved fracture gradient methodology-understanding the minimum stress in Gulf of Mexico. Paper ARMA 10-177 presented at the 44th U.S. Rock Mechanics Symposium and 5th U.S.-Canada Rock Mechanics Symposium, Salt Lake City, UT, USA, 27–30 June, 2010.

Maia, A., Poilate, C.E., Falcao, J.L. and Coelho, L.F.M.: Triaxial creep tests in salt applied in drilling through thick salt layers in Campos Basin-Brazil. Paper SPE/IADC 92629 presented at the SPE/IADC Drilling Conference, Amsterdam, The Netherlands, 23–25 February, 2005.

Marinez, R., Shabrawy, M.E., Sanad, O. and Waheed, A.: Successful primary cementing of high pressure salt water kick zones. Paper SPE 112382 presented at the SPE North Africa Technical Conference & Exhibition, Marrakech, Morocco, 12–14 March, 2008.

Shen, X.P.: DEA-161 Joint Industry Project to Develop an Improved Methodology for Wellbore Stability Prediction: Deepwater Gulf of Mexico Viosca Knoll 989 Field Area. Halliburton Consulting, Houston, TX, USA, 18 August, 2009.

Shen, X.P., Bai, M. and Smith, R.: Numerical analysis on subsalt wellbore stability in the Viosca Knoll deepwater of the northern region of the Gulf of Mexico. Paper SPE 130635 presented at the CPS/SPE International Oil & Gas Conference and Exhibition in China, Beijing, China, 8–10 June, 2010.

Shen, X.P., Diaz, A. and Sheehy, T.: A case study on mud-weight design with finite-element method for subsalt sells. Paper 1120101214087 accepted by and to be presented at the 2011 ICCES, Nanjing, China, 18–21 April, 2011.

Zoback, M.D.: *Reservoir Geomechanics*. Cambridge University Press. New York, NY, USA, 2007.

CHAPTER 5

Mud weight design for horizontal wells in shallow loose sand reservoir with the finite element method

Xinpu Shen, Sakalima Sikaneta & Johan Ramadhin

5.1 INTRODUCTION

Conventional mud weight designs are typically developed by using one-dimensional (1D) analytical solutions. These methods, however, are unreliable in shallow, unconsolidated sands, which have very low cohesive strength and pressure-sensitive material properties, such as Young's modulus and Poisson's ratio. Using a 1D solution for these reservoirs can result in unnecessarily high mud weights and narrow mud weight windows, especially in deviated wells. In this case, a numerical method with a pressure-dependent material model becomes necessary for predicting a minimum safe mud weight. A fully coupled, nonlinear three-dimensional (3D) finite element model (FEM) was built for a shallow 305 m true vertical depth (TVD) horizontal well in the Tambaredjo NW field of Suriname. The calculation of deformation of loose sand, which has pressure-dependent material properties, including cohesive strength, internal frictional coefficient, and Young's modulus, was combined with porous flow. We used this model to investigate wellbore stability and the minimum safe mud weight gradient. Core samples from wells in the Tambaredjo NW field of Suriname were analyzed. The results of the analyses were used to develop a nonlinear relationship between material parameters, including cohesive strength, internal frictional coefficient, Young's modulus, and the mean stress in effective stress space. The pattern of variation of these pressure sensitive parameters and other data were drawn from the multiple disciplines involved in the field development plan. The results of the study show the model-predicted minimum mud weight gradient required to drill the well without instability can differ from the conventional analytical solution by as much as 2348.6 Pa/m (2 ppg). The use of this pressure-dependent material model for predicting a minimum safe mud weight in shallow, unconsolidated sand reservoirs can result in significant savings in field development costs.

The Tambaredjo and Tambaredjo NW fields are located in the Guyana basin along the northern margin of South America between the Amazon and Orinoco deltas, and immediately north of the Guyana Shield (see Figure 5.1). The reservoirs in the field consist of unconsolidated Tertiary sands that were deposited in an estuarine environment. Hydrocarbons are thought to be sourced from the Cretaceous Canje formation located down-dip and offshore. The fields are characterized by a gently northward dipping monoclinal structure.

The rock strength in the field is deemed to be low. In particular, sandy formations, such as the reservoir intervals, have little to no cohesion. Characterizing the strength within these lithologies is particularly difficult. The large grain size and high porosity of the reservoir formations prevents capillary strengthening. Based on an analytical capillarity analysis, increasing water cut with the production of hydrocarbons is not expected to significantly alter the formation strength. The low shear strength of the sands mandates sand production prevention equipment, such as screening and gravel packing.

Particular emphasis was placed on analyzing the feasibility of drilling horizontal wells in the field. The optimal direction to prevent shear-induced borehole collapse in horizontal wells is in the north-south direction, whereas the optimal direction to produce shear-enhanced porosity in horizontal wells is in the east-west direction.

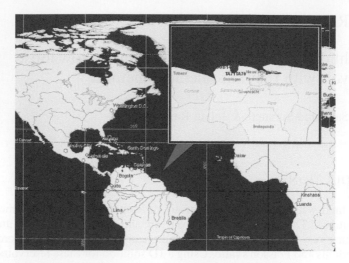

Figure 5.1.   Project map showing the location of the study area.

Many researchers have performed numerical studies of the mechanical behavior of reservoir sand (see Boström 2009; Han and Dusseault 2005; Yi *et al.*, 2008). The investigation of unconsolidated heavy-oil sand has drawn the attention of several researchers in recent years (Ahmed *et al.*, 2009; Hilbert *et al.*, 2009). Recently, the integrated use of 2D analytical software with 3D numerical software to simulate the mechanical behavior of the reservoir has been reported (Shen 2009). It was shown that the 3D FEM can properly account for the mean-stress dependent properties of Young's modulus and Poisson's ratio. Pore pressure results predicted by 2D analytical software with logging data can be used as input data for the FEM.

This chapter includes an overview of the geological factors affecting the area, as well as the sources of data made available for the study and a review of the quality and use of the data. It also provides analyses of the approach used to determine the pre-drill in-situ stresses and the strength and material behavior of the formations of interest. The chapter also presents the results of the numerical models used to determine the minimum mud weight required for drilling directional and, in particular, horizontal wells; it concludes with a brief discussion and recommendations.

5.2   GEOLOGICAL SETTING AND GEOLOGICAL FACTORS AFFECTING GEOMECHANICS

The geometric controls of reservoir distribution in the Tambaredjo fields are largely sedimentological. Migrating facies associated with the estuarine environment determine the depositional sites of sandy and clayey sediments. Relative changes in sea level overprint this lateral variability.

The lack of cohesion of the reservoir sands restricts the use of standard linear elastic models in the prediction of the geomechanical behavior of the reservoir sands without significant calibration data. The relatively young age and shallow depth of the sands implies that close-packing, interlocking, and overgrowth mechanisms that strengthen deeper sands may not be operative. Consequently, the sands are expected to be weak and to have low cohesion. This interpretation is confirmed by low recovery rates for cores and low coefficients of friction in triaxial test samples.

The Tambaredjo fields are located on a relatively poorly studied, tectonically passive margin. The closest active, large-scale tectonic features are the strike slip and compressional zones of northern Venezuela and Trinidad to the west, the obducting Caribbean plate to the northwest (with the Atlantic Oceanic plate being subducted), and the extensional spreading Mid-Atlantic Ridge to the east. Within the geological regional setting of a passive margin with a shelf oriented east-west, the most likely stress field is that the maximum horizontal stress strikes east-west. The vertical stress is greater than both the horizontal stresses in this region. The two horizontal stresses are relatively similar in magnitude.

Geological analogs can provide valuable information for projects in which the data is limited. The shallow heavy oil fields of the western Canadian sedimentary basin and the Faja del Orinoco in Venezuela provide valuable analogs for the Tambaredjo fields. The key differences between the Canadian setting and the Tambaredjo fields are the stress fields (compressional vs. passive) and the age (old vs. young). Shallow, low-cohesion reservoirs in the Faja del Orinoco likely provide the best analog for Tambaredjo, although the Faja del Orinoco reservoirs tend to be deeper.

5.3 PORE PRESSURE AND INITIAL GEOSTRESS FIELD: PREDICTION MADE WITH LOGGING DATA AND ONE-DIMENSIONAL SOFTWARE

5.3.1 *Pore pressure*

The depth and age of the sediments of the Tambaredjo field, as well as the drilling reports, imply that the pore pressure regime is likely to be normal. This implication is confirmed by log data from 40 wells that were analyzed by using normal compaction trend analysis with the Drillworks® set of 1D analytical software.

Normal compaction trends in both resistivity and sonic datasets were determined. The analysis based on these trends indicates that a normally pressured regime is present in the field area. A freshwater effect shown in the resistivity data indicates that the sonic dataset is more reliable for pore pressure prediction.

The pore pressure analysis was based only on trends observable in shale lithologies. Total gamma ray or potassium-thorium gamma ray (when available) was used to discriminate between the shale and non-shale lithologies.

The compaction trend in the resistivity datasets was analyzed using the deep resistivity curve and Eaton's normal compaction trend. The compaction trend used for the sonic datasets was based on Bower's normal sonic compaction trend. Consistent regional parameters were applied for the pore pressure analysis because of the similarity between well logs. The values of A and B adopted for Bower's method were 1.75 and 1.15, respectively. The mudline sonic value that was used was 6.5617 s/m (2,000 ms/ft). For Eaton's resistivity pore pressure method, an exponent of 0.6 was applied.

5.3.2 *Stress field orientation*

Both the magnitude and the direction of the stress tensors control the wellbore stability and geomechanical performance. Usually, in a region where tectonic stress is not significant, it can be assumed that the vertical stress is one of the principal stresses and that the orientation of the horizontal stresses are therefore the only other directional components of the stress field that must be defined. The World Stress Map (2008) does not indicate the stress regime in the vicinity of the Tambaredjo fields. However, moment tensors for earthquakes recorded by the United States Geological Survey Earthquakes Hazards program (2009) support the interpretation that the maximum horizontal stress orientation strikes in an east-west direction. An earthquake recorded at 9.56° N, −59.160° E on September 24, 1989 registered slip of +167 mm and +17 mm on doubly coupled planes oriented with strikes of 14° N and 108° N, and with dips of 74° and 78°, respectively. An earthquake recorded at 4.88° N, −51.96° E on June 8, 2008 registered slip of −162 mm and −19 mm on doubly coupled planes oriented with strikes of 273° N and 178° N, and dips of 72° and 73° respectively. This information about earthquake focal mechanisms provides an estimate of the regional stress field orientation, but not the magnitude of stresses.

5.3.3   *Overburden gradient (vertical in-situ stress)*

In passive margin settings, the vertical stress is the largest principal stress. The overburden gradient corresponding to the vertical stress was calculated for the 40 wells for which log data was provided by extrapolating the density log to the surface and integrating with respect to the Kelly bushing referenced depth. The resultant overburden gradient indicates that, in the vicinity of the reservoirs (~304.8 m), the overburden gradient is approximately 17027.35 Pa/m (14.5 ppg).

5.3.4   *Minimum in-situ stress*

The minimum in-situ stress is expected to be horizontal and to be oriented northward, based upon the tectonic setting of the project area. A Matthews and Kelley type relationship $\sigma_h^{eff} = K_h \sigma_v^{eff}$ is used to estimate the minimum horizontal stress. Typically, K_h is approximately 0.8 in passive margin settings. A leakoff test conducted in well 29Ow19 at approximately 880 ft (267.00 m) indicated a minimum horizontal stress of 150 psi. This corresponds to an equivalent mud weight of:

$$EMW = MW + (\sigma_h/0.5195 * TVD) = 12.3\,\text{ppg} = 14443.8\,\text{Pa/m} \qquad (5.1)$$

At a depth of 880 ft (268.2 m), the overburden gradient is approximately 15.86 ppg (18624 Pa/m) (average of values from all available wells), which results in a K_h value of 0.775. This value is used in the calculations of the minimum horizontal stress for the 40 wells analyzed.

5.3.5   *Maximum in-situ horizontal stress*

The maximum *in-situ* stress is expected to be horizontal and to be oriented east-west, based upon the tectonic setting. A relationship of the form is given as follows:

$$\sigma_H = \sigma_h + T_f\left(\sigma_v - \sigma_h\right) \qquad (5.2)$$

Eq. 5.1 is used to predict the maximum horizontal stress. For passive margin settings, T_f varies between zero and one. Image logs showing borehole breakouts can be used to constrain the value of T_f if borehole breakouts are evident. The image logs available for this project (obtained for wells #29Og06, #29Om17, and #29Ut02) do not show breakouts; consequently, this type of analysis could not be performed. However, the structural setting of the Tambaredjo field implies a relatively isotropic horizontal stress field; thus, T_f has a relatively low value. Low T_f values depress the shear fracture gradient, and this effect is noted in the Tambaredjo wells. A T_f value of 0.25 was chosen to reflect the low level of horizontal stress anisotropy. As previously described, regional tectonic forces indicate that the maximum horizontal stress is oriented east-west.

5.4   FORMATION STRENGTH AND GEOMECHANICAL PROPERTIES

The strength of geological formations is not a single number; rather, it is a complex function of the type of geological material under consideration. For most formations, two numbers usually suffice to describe the strength of the rock. Any two of the parameters (unconfined compressive strength (UCS), cohesive strength (CS), friction angle (f), or coefficient of friction (m)), can be used to define two-parameter rock strength models.

Formation strength parameters are related to formation slowness, as are elastic properties. To estimate the strength of the formations in the Tambaredjo fields, Horsrud's correlations for cohesive strength and friction angle were used. The Horsrud correlations, which were developed for North Sea Tertiary shales, yield effective friction angle and cohesive strength values that are reflective of the shale intervals only.

Certain wells did not contain sonic data from which strengths could be derived for geomechanical modeling. For these wells, a Gardner transform was used to derive a synthetic slowness log from density. The Gardner transform is given in Eq. 5.2 as:

$$DT = 10^6 \left(\frac{RHOB}{A} \right)^{-1/B}$$
(5.3)

where DT is slowness, $RHOB$ is bulk density, and A and B are empirically-derived coefficients. The value of A used in the analyses was 0.23, and the value of B was 0.25. These values reproduce synthetic sonic logs that are similar to measured sonic logs.

Particular emphasis was placed on the feasibility of drilling horizontal wells in the field development plan. Typically, for cohesionless sands, such as the Tambaredjo reservoirs, more than two parameters are needed to describe formation strength and deformation. To model these formations, their behavior over a range of pressures must be determined experimentally, or estimated based upon analog settings. The results of the triaxial test data from well #1M10.1, shown in Table 5.1, highlight the variability in elastic properties over a range of confining stresses. Figure 5.2 and Figure 5.3 show the curves of this mean-stress dependency for Young's modulus and Poisson's ratio. The triaxial tests measured a cohesive strength of 11.089 MPa for a sample taken at 290 m TVD. This value is deemed to be disproportionably high, and it is believed that the actual cohesive strength value is much closer to zero. Unconsolidated sands have effectively zero cohesion, and it is unlikely that the sands will have any additional strength imparted from capillary effects. The high measured cohesion value likely reflects sampling bias toward more competent parts of the formation that was cored. Table 5.2 lists the compressive strength and Mohr-Coulomb parameters from a triaxial test of cores from borehole #1M101.

A computer program based upon the analytical model of Detournay and John (1988) was developed to determine the strength of capillary forces in the reservoir section and the influence of water cut on the capillary strength. Data entered for this custom-developed program includes grain size diameter, porosity, and surface tension at the oil/water interface. The effective grain size was determined using the relation in Eq. 5.3 as:

$$D_{eff} = \frac{\sum f_i d_i^3}{\sum f_i d_i^2}$$
(5.4)

where the d_i are diameters of grain size fraction f_i. Using the available grain size distribution data, an effective grain size of 0.3 mm was determined. A sensitivity analysis was performed

Table 5.1.   Triaxial static Young's modulus, poisson's ratio, and compressive strength.

Sample number	Depth (m)	Confining pressure (MPa)	Bulk density (kg/m³ × 10³)	Compressive strength (MPa)	Young's modulus (MPa)	Poisson's ratio
28	297.18	0.67	2.13	4.81	206.84	0.13
29	297.24	8.27	2.13	20.99	1585.8	0.24
30	297.30	2.76	2.14	8.87	344.74	0.18

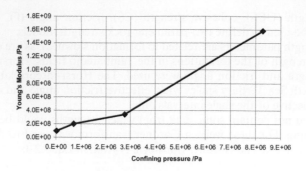

Figure 5.2.   Variation of Young's modulus with confining pressure.

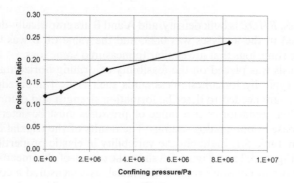

Figure 5.3.   Poisson's ratio with confining pressure.

Table 5.2.   Compressive strength and Mohr-Coulomb parameters from triaxial test on cores from borehole #1M10.1.

Sample depth (m)	Confining pressure $P_c = \sigma_3$ (MPa)	Compressive strength σ_1 (MPa)	Slope on σ_1 vs. σ_3	Unconfined compressive strength (MPa)	Cohesion (MPa)	Angle of internal friction
297.18	0.69	4.81				
297.18	2.76	8.87				
297.18	5.52	15.86	2.19	3.23	1.09	21.88

for a range of surface tension, friction angle, and porosity, using values in the range of those reported in the relative permeability tests, and the fluids analysis results provided by Staatsolie. Within all likely ranges of these parameters, capillary forces were found to be low at all saturations. Consequently, the sandy sections of the reservoir are not likely to change dramatically in strength as a result of water cut.

These findings presented indicate that the Tambaredjo NW field is hydrostatically pressured, and that conventional wells, such as those for which data were furnished, are not susceptible to wellbore instability problems while drilling. The ratio of the vertical to minimum horizontal in-situ effective stress is near 0.775. It is likely that the in-situ horizontal stresses are similar in magnitude based upon the geological setting. The orientation of the maximum horizontal stress is interpreted to be east-west. Based on this interpretation of the stress field, planes of shear induced porosity enhancement should be aligned to strike east-west and dip at approximately 60° to the north (or south). To reduce the risk of shear collapse while drilling, the optimal direction to drill horizontal wells is in the north or south directions.

Because the shale formations are more competent than the sands, it is recommended that kick off points be located in shaly sections, if possible. Caliper data from the offset well indicate a strong grain size effect that is not captured in the models presented in this chapter. To enhance wellbore stability, drilling with slimhole or coiled tubing would be particularly beneficial.

5.5   FINITE ELEMENT MODEL

The geostress analyses made with Drillworks software are based upon a linear-elastic model of the formation and cannot account for the pressure sensitive nature of the sands. A fully coupled non-linear elasto-plastic model was constructed to better assess the geomechanical factors affecting horizontal drilling. The FEM that was constructed also accounts for changes in pore pressure as a result of deformation; it can also model plastic deformation processes. Plastic deformation modeling is critically important in low cohesion sands because these materials can "plastically flow" at very low stresses.

Figure 5.4 shows the FEM geometry. A section of horizontal borehole has been taken as the objective for numerical simulation. Length, width, and height are given as 3 m, 3 m, and 1 m, respectively. As shown in Figure 5.5, all boundaries are permeable except the borehole wall. A Kozney-Carman relationship was used to define changes in permeability with porosity during loading, using porosity values extracted from permeability test data. The pressure-Young's modulus and pressure-Poisson's ratio relationship from the triaxial test data of 1M10.1 (Figure 5.2 and Figure 5.3) was extrapolated to add one additional point at low confining pressure. All other material parameters were derived from core data and test

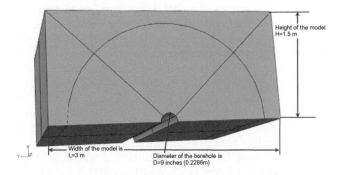

Figure 5.4.   Model geometry. Length in the wellbore-axial direction is 3 m.

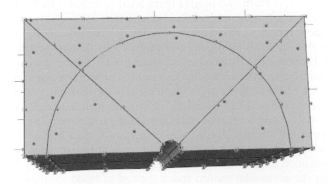

Figure 5.5.   Model loading and boundary conditions.

data. A Mohr-Coulomb failure criterion was adopted for simplicity; therefore, the model results may be considered to be conservative. Note that the units used in this analysis are in MPa rather than in psi because those are the units used in the analysis program. The conversion from psi to MPa is that 1 psi = 0.0068947 MPa.

The loading parameters were chosen to reflect the least favorable orientation of a horizontal well at 305 m TVD. Loading is described as follows: the vertical stress was 5.8 MPa, minimum horizontal stress was 5.0 MPa, and the maximum horizontal stress (parallel to borehole) was 5.2 MPa. The pore pressure was 3.13 MPa. Several different simulations were run using various values of mud weight and cohesive strength. The cohesive strength of the reservoir sands is the most dominant factor in constraining the geomechanical behavior of the reservoir. The value measured in the laboratory was deemed to be too high, and a conservative value of 0.01 MPa was used to develop a worst-case scenario. The friction angle used in the analysis was 21.88°, reflecting the value determined by the triaxial test results, and additional models were run using a more likely in-situ value of 25°. A dilation angle of 18° was used to model post yield deformation, and the initial void ratio of the sand was assumed to be 0.53.

As shown in Figure 5.5, surface traction is applied on the top surface. The in-situ stress field is applied at all points in the interior of the model. The mud weight applies a load on the surface of the borehole after it has been drilled. The four lateral sides and the bottom are constrained with zero displacement conditions. Pore pressure boundary conditions are applied at all sides except the symmetric surface, which is applied at the bottom of the model.

5.6 NUMERICAL RESULTS WITH FINITE ELEMENT MODELING

This section describes the results of model simulations in which the mud weight and friction angle are varied. As expected, as the mud weight and friction angle are increased, the plastic yield zone decreases in size, as does the deformation. Boundary effects are evident at the edge of the models, and the most representative cross section of the model is in the center. With respect to wellbore stability, it is a rule of thumb to avoid mud weights that will result in the entire wellbore being surrounded by failed material. Based on this design criterion, the minimum mud weight should be 12330.1 Pa/m (10.5 ppg). Note that the figures that illustrate total displacement do not take creep processes into account. Because the Tambaredjo wells are drilled very fast, creep is not expected to play as significant a role in deformation as would be the case in wellbores that remain in the openhole state for extended periods of time.

Figure 5.6 through Figure 5.12 show the numerical results at 587.2 Pa/m (0.5 ppg) increments, from 11155.8 Pa/m (9.5 ppg) (Figure 5.6) to 14091.5 Pa/m (12 ppg) (Figure 5.12). The internal friction angle for all of these results is held constant at 21.88°.

Figure 5.13 through Figure 5.18 show the model results at 587.2 Pa/m (0.5 ppg) increments, from 11155.8 Pa/m (9.5 ppg) (Figure 5.13) to 13504.4 Pa/m (11.5 ppg) (Figure 5.18). The friction angle for these results is held constant at 25°.

Elasto-plastic modeling based upon rock strength parameters from Horsrud's correlations (Sikaneta and Shen 2009) yield shear failure gradients that are relatively reasonable for shale formations. However, the Horsrud correlations fail to capture the cohesionless nature of the sands; analytical elastoplastic models also fail to capture the essential features of sand geomechanics. A FEM was developed to better assess the feasibility of drilling horizontal wells at approximately 305 m. TVD. Based on the finite element and analytical models, a minimum mud weight of 10.5 ppg is recommended for drilling a horizontal well at 305 m. The relatively low pressure in the shallow Tambaredjo reservoirs should help to mitigate the risk of differential sticking. The fracture gradient at 305 m is approximately 17027.3 Pa/m (14.5 ppg). The recommended mud window for drilling is between 12330.1 to 17027.3 Pa/m (10.5 and 14.5 ppg) for the horizontal section and at the end of build sections.

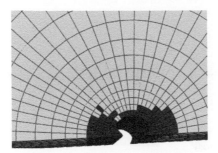

Figure 5.6.   The resultant active plastic zone distribution around the wellbore at PMW = 10568 Pa/m (9 ppg), Φ = 21.88°.

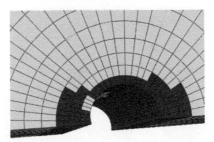

Figure 5.7.   The resultant active plastic zone distribution around the wellbore at PMW = 11155.8 Pa/m (9.5 ppg), Φ = 21.88°.

Figure 5.8.   The resultant active plastic zone distribution around the wellbore at PMW = 11743 Pa/m (10 ppg), Φ = 21.88°.

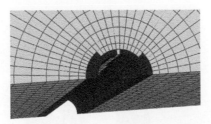

Figure 5.9.   The resultant active plastic zone distribution around the wellbore at PMW = 12330 Pa/m (10.5 ppg), Φ = 21.88°.

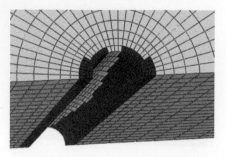

Figure 5.10.   The resultant active plastic zone distribution around the wellbore at PMW = 12917.2
Pa/m (11 ppg), Φ = 21.88°.

Figure 5.11.   The resultant active plastic zone distribution around the wellbore at PMW = 13504.4
Pa/m (11.5 ppg), Φ = 21.88°.

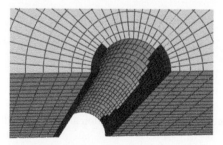

Figure 5.12.   The resultant active plastic zone distribution around the wellbore at PMW = 14091.5
Pa/m (12 ppg), Φ = 21.88°.

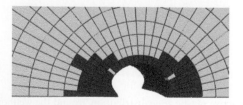

Figure 5.13.   The resultant active plastic zone distribution around the wellbore at PMW = 10568.7
Pa/m (9 ppg), Φ = 25°.

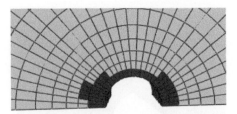

Figure 5.14. The resultant active plastic zone distribution around the wellbore at PMW = 11155.8 Pa/m (9.5 ppg), Φ = 25°.

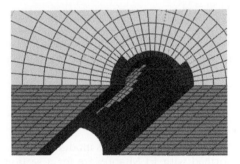

Figure 5.15. The resultant active plastic zone distribution around the wellbore at PMW = 11743 Pa/m (10 ppg), Φ = 25°.

Figure 5.16. The resultant active plastic zone distribution around the wellbore at PMW = 12330.1 Pa/m (10.5 ppg), Φ = 25°.

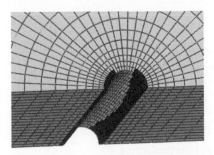

Figure 5.17. The resultant active plastic zone distribution around the wellbore at PMW = 12917.2 Pa/m (11 ppg), Φ = 25°.

Figure 5.18. The resultant active plastic zone distribution around the wellbore at PMW = 13504.4 Pa/m (11.5 ppg), $\Phi = 25°$.

Based on an analytical model of the effect of water cut on cohesive strength, there is little likelihood of production-induced saturation changes causing problematic weakening of unconsolidated sand.

5.7 CONCLUSIONS

With the mean-stress dependent model for Young's modulus and Poisson's ratio, 3D numerical analysis has been performed with FEM software for a section of horizontal wellbore within a loose sand reservoir. Experimental values of cohesive strength parameters at the target depth have been presented. Geomechanical parameters and pore pressure were predicted with 1D analytical software by using logging data, which provided a sound basis for numerical analysis.

The results of minimum safe mud weight values predicted here were successfully used in a field development plan for the Tambaredjo NW field in Suriname. This case study provides a good example of the integrated use of 2D prediction software with 3D FEM numerical software.

ACKNOWLEDGEMENTS

The authors would like to thank Alejandro Arboleda, project manager from Halliburton Consulting—Geomechanics Practice Group, for his management over the project related to this chapter, as well as Franklin Sanchez and Jose Luis Ortiz Volcan, Halliburton regional managers, and Lilian Mwakipesile-Arnon of Staatsolie for their assistance with the works related to this chapter.

NOMENCLATURE

A	=	Model parameter
B	=	Model parameter
c	=	Cohesive strength, Pa
d_i	=	Diameters of grain size fraction, f_i, m
D_{eff}	=	Effective grain size, m
σ_h^{eff}	=	Minimum horizontal effective vertical stress, Pa
σ_v^{eff}	=	Vertical effective vertical stress, Pa
K_h	=	Effective stress ratio
T_f	=	Tectonic factor

σ_v = Effective vertical stress gradient, Pa
σ_H = Maximum horizontal effective vertical stress, Pa
ϕ = Internal friction angle, °
ρ = Density, kg/m³
CS = Cohesive strength
DT = Sonic data
EMW = Mud weight equivalent, Pa/m (ppg)
FEM = Finite Element method
MW = Mud weight, Pa/m (ppg)
RHOB = Bulk density kg/m3
TVD = True vertical depth, m (ft)
UCS = Unconfined compressive strength, Pa

REFERENCES

Ahmed, K., Khan, K. and Mohamad-Hussein, M.A.: Prediction of wellbore stability using 3D finite element model in a shallow unconsolidated heavy-oil sand in a Kuwait field. Paper SPE 120219-MS presented at the SPE Middle East Oil and Gas Show and Conference, Bahrain, Bahrain, 15–18 March, 2009.

Bostrøm, B.: Development of a geomechanical reservoir modelling workflow and simulations. Paper SPE 124307 presented at the SPE Annual Technical Conference and Exhibition, New Orleans, LA, USA, 4–7 October, 2009.

Detournay, E. and St. John, C.M.: Design charts for a deep circular tunnel under non-uniform loading. *Rock Mechanics and Rock Engineering* 21 (1988), pp. 119–137.

Han, G. and Dusseault, M.: Sand stress analysis around a producing wellbore with a simplified capillarity model. *Int. J. Rock Mech. Mining Sci.* 42 (2005), pp. 1015–1027.

Hilbert Jr., L.B., Saraf, V.K., Birbiglia, D.K.J., Shumilak, E.E., Schutjens, P.M.T.M., Hindriks, C.O.H. and Klever, F.J.: Modeling horizontal completion deformations in a deepwater unconsolidated shallow sand reservoir. Paper SPE 124350-MS presented at the SPE Annual Technical Conference and Exhibition, New Orleans, LA, USA, 4–7 October, 2009.

Khaksar, A., Rahman, K., Ghani, J. and Mangor, H.: Integrated geomechanical study for hole stability, sanding potential and completion selection: a case study from southeast Asia. Paper SPE 115915-MS presented at the SPE Annual Technical Conference and Exhibition, Denver, CO, USA, 21–24 September, 2008.

Shen, X.P.: DEA-161 *Joint Industry Project to Develop an Improved Methodology for Wellbore Stability Prediction: Deepwater Gulf of Mexico Viosca Knoll 989 Field Area.* Halliburton Consulting, Houston, TX, USA, 18 August, 2009.

Sikaneta, S. and Shen, X.: *Pore Pressure and Wellbore Stability Prediction in the Tambaredjo Field, Suriname.* Halliburton Consulting, Houston, TX, USA, 3 March, 2009.

Yi, X., Goodman, H.E., Williams, R.S. and Hilarides, W.K.: Building a geomechanical model for Kota-batak field with applications to sanding onset and wellbore stability predictions. Paper IADC/SPE 114697-MS presented at the IADC/SPE Asia Pacific Drilling Technology Conference and Exhibition, Jakarta, Indonesia, 25–27 August, 2008.

United States Geological Survey Earthquakes Hazards Program. 2009. www.earthquake.usgs.gov/regional/neic. Downloaded 19 March 2009.

World Stress Map. 2008. http://www-wsm.physik.uni-karlsruhe.de. Downloaded 19 March 2009.

CHAPTER 6

A case study of mud weight design with finite element method for subsalt wells

Xinpu Shen, Arturo Diaz & Timothy Sheehy

6.1 INTRODUCTION

During the last decade, a significant effort has been made to explore and develop large oil and gas reservoirs associated with salt structures (Dusseault *et al.*, 2004). Examples of the most active areas of subsalt drilling include the following:

- Jurassic salt emplaced during tertiary (Gulf of Mexico).
- Late Pennsylvanian to Early Triassic anticlines (Paradox basin).
- Mid US continent Devonian Age (Williston basin).
- Zechstien-age salt emplaced in the Cretaceous (North Sea).
- Zagros salt plugs (Iran).
- Brazilian and West Africa offshore basins.
- Kashagan and Tengiz oilfields in the Caspian basin in Kazakhstan.

These examples range from 3 to 7 km of salt body, with a total depth of more than 8.3 km. The geometries of the salt structures vary and include domes, ridges, salt tongues, pillows, undeformed bedded sedimentary salts, and mixed domains. As a result of the economical successes of subsalt reservoirs, a significant amount of exploration and new field developments will continue in salt provinces over the next decade, leading to challenges in both the design and drilling of wells in these complex geological environments (Tsai *et al.*, 1987).

As described by Whitson and McFadyen (2001), drilling in salt provinces give rise to two fundamental issues that require careful management of mud weight: 1) the rate of salt creep when drilling into the salt body, and 2) the determination of pore pressure and fracture gradient in both non-salt formations interbedded in the salt and when exiting the salt.

The geometry of a salt body affects the drilling strategy. In some cases, it can limit the well trajectory (salt tongues and bedded salts); in other cases, it can expand the well trajectory possibilities (diapiric salt) to avoid the salt or to limit the drilling of as much salt as possible, according to the geomechanical state of the rock surrounding the salt, depth, and creep rate of the salt.

Because salt is tectonically mobilized by the density difference between salt ($2.16 \times 10^3 \, \text{kg/m}^3$) and the other sediments (2.3–$2.6 \times 10^3 \, \text{kg/m}^3$) as a result of its viscous behavior at moderate load and temperature, the stress field in the surrounding formations is perturbed. This perturbation leads to stress rotation near the salt-clastic interface and to change in stress magnitude that can differ significantly from the far field stresses. As a result, subsalt overpressure or pressure reversion may exist and create extensive rubble or highly-sheared zones below or adjacent to the salt diapirs. This is an intrinsic consequence of the equilibrium stress field needed to satisfy the various stresses that exist within the salt body and the adjacent formation; it causes extremely expensive and difficult wellbore stability problems during drilling, completion, or production, and represents a challenge for the oil industry in both operations and modeling.

Salt affects the actual geomechanical environment through the alteration of the local stresses because it cannot sustain deviatoric stress. Consequently, it deforms by means of plastic creep in response to any imposed deviatoric stress. Changes in the stress field and shear stresses adjacent to the salt may be sufficient to cause a reorientation of the principal stress.

As previously mentioned, fundamental wellbore stability problems during drilling in salt environments consist of two critical sections: while drilling through the salt and when exiting the salt.

Wellbore stability problems while drilling through salt are easier to control by managing the density and temperature of drilling fluid. This control is possible because salt behaves as a highly viscous fluid with a creep rate that strongly depends on temperature and stress difference. Salt is impermeable and geochemically inert except in high solubility (easily managed with a saturated water phase in the mud). In addition, salt has a tensile strength (unfractured) that is more reliable than porous formations. Because no pore pressure is present, we can increase the wellbore pressure to near the lithostatic pressure inside of salt without any problem. However, caution should be observed because of the presence of shale intrusion or interbedded bischofite and carnallite. At this stage, increasing mud density, maximizing rate of penetration, oversaturating the water phase of the fluid, and cooling the mud are the most common strategies (Whitson and McFadyen 2001) to control the closure rate and washout of the well.

Exiting the salt is the most critical stage in drilling because of two possible occurrences: a pressure and stress reversion zone leading to a sudden lost-circulation problem, or trapped overpressure zones leading to a blowout. Consequently, the problem is difficult to anticipate because the presence of salt impairs the pore pressure and fracture gradient prediction based on velocity and wireline logging data.

The high uncertainty in the pore pressure prediction and the presence of rubble and highly-sheared zones require accurate predictions of subsalt formation velocities using the most up-to-date techniques in seismic imaging along with numerical simulation to anticipate the appropriate mud weight window (MWW) and to minimize subsalt losses (Power et al., 2003).

Fredrich et al. (2003) reported the results of a three-dimensional (3D) finite element method (FEM) simulation of various geometries of salt bodies, including spherical salt bodies, horizontal salt sheets, columnar salt diapers, and columnar salt diapers with an overlaying tongue. Their results include the following:

- The shear stress is highly amplified in certain zones in specific geometries adjacent to the salt.
- The salt body induces anisotropy of horizontal stresses.
- The principal stresses are no longer vertical and horizontal, and the vertical stress may not be the maximum stress.
- For some geometries, vertical stress within and adjacent to the salt is not equal to the gravitational load (stress-arching effect occurs) before drilling or production and behaves similarly to the overburden changes during depletion of compactable reservoirs.
- The stress perturbation can only be determined by solving the complete set of equilibrium, compatibility, and constitutive equations with the appropriate initial and boundary conditions using a FEM method.

Cullen et al. (2010) emphasized the importance of 3D FEM modeling of a subsalt region and discussed the effect of recent technological advances in seismic imaging, pore pressure prediction, and geomechanical modeling to improve drilling efficiency when exiting the salt.

Because of the large differences between the field scale and casing-section scale in modeling, it has been difficult, if not impossible, to combine these models in the past. Existing examples of numerical analysis of casing failure were either performed at reservoir scale without direct coupling to behaviors at the field scale, or performed at a much larger scale, which sacrificed much needed modeling resolution.

Submodeling techniques are used to accommodate the field-to-casing-section scale discrepancy. The submodeling technique includes using a large-scale global model to produce boundary conditions for a smaller-scale submodel. In this way, the hierarchical levels of the submodel are not limited.

It should be noted that calculation presented in this chapter is illustrative, and all data are either fictitious or modified on the basis of some practical cases in engineering. No data is directly from any commercial project.

The contents presented in the following sections include the following:

- A brief introduction of some essential concepts used in MWW design
- Global model description and numerical results
- Model description of submodels and numerical results obtained with submodels
- Results obtained with a one-dimensional (1D) solution tool
- Analysis of effective stress ratio in terms numerical results of 3D FEM
- Comparison between results of the MWW obtained the 3D FEM and those obtained with 1D method
- Conclusions

6.2   BRIEF REVIEW OF CONCEPTS OF MWW AND NUMERICAL PROCEDURE FOR ITS 3D SOLUTION

6.2.1   *Brief review of mud weight window concepts*

The MWW is the range of mud density values that provides safe support to the wellbore during the drilling process at a given depth. If the mud weight value is selected within the MWW range, the wellbore is stable, and no plastic deformation will occur on the wellbore surface. Furthermore, with a safe mud weight selected within the MWW, no mud loss will occur. Therefore, the MWW is defined by two bounds. The lower bound is the shear failure gradient (SFG), which is the minimum mud weight required to avoid plastic failure of the wellbore. The upper bound is the fracture gradient (FG), which is the maximum value of mud weight that cannot induce any fracture opening. Because natural fractures usually exist within various kinds of formations and wellbores are primarily vertical, in practice, the value of the minimum horizontal stress is taken as the value of FG.

In practice, the MWW of a given wellbore can be designed with either a 1D analytical method or a 3D numerical FEM. The 1D method determines the horizontal stress components in terms of overburden stress and logging data along the wellbore trajectory, and only the information along the wellbore trajectory is used in the MWW determination. As a type of 3D method, the FEM uses a 3D model that consists of 3D geometry and a 3D mechanical constitutive relationship.

The input data required for the 1D prediction of MWW consists of two parts:

- Pore pressure (PP), overburden gradient (OBG), and effective stress ratio and/or Poisson's ratio
- Cohesive strength (CS), friction angle (FA) and/or unconfined compressive strength (UCS), and tectonic factor

The first part of the input data is used to predict the upper bound of MWW, which is the FG. The second part the of input data is used to predict the lower bound of MWW, which is the SFG. Among these data, the effective stress ratio is used to calculate the minimum horizontal stress (also regarded as FG), and the tectonic factor is used to calculate the maximum horizontal stress. Poisson's ratio can be used instead of the effective stress ratio. The Drillworks application will calculate the effective stress ratio in terms of Poisson's ratio.

Eq. 6.1 defines the effective stress ratio k_0:

$$k_0 = \frac{S_h - pp}{OBG - pp} \tag{6.1}$$

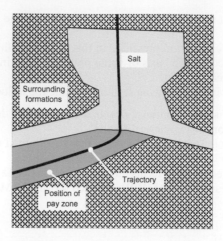

Figure 6.1.   Sectional view of geostructure of the field.

where S_h is minimum horizontal stress. Figure 6.1 illustrates the trajectory of the wellbore and the formations. It is a vertical well through a salt body. The thickness of salt body where the wellbore penetrates the reservoir is 5600 m. The width of the model built in the calculation is 8000 m, and the height is 9000 m. The target formation is at the salt base at the true vertical depth (TVD) interval of 7500 to 8500 m. The MWW must be predicted at this TVD interval. Values of material parameters are given in Table 1. With reference to the data and suggestion in Mitchell and Mouchet (1989), values of effective stress ratio k is calculated in terms of Poisson's ratio v:

$$k = \frac{v}{1-v} \tag{6.2}$$

With the given values of Poisson's ratio in Table 6.1 in the next section, the effective stress ratio can be obtained as 0.43.

The tectonic factor is another type of stress-related input data. It is used to determine the SFG, which is the lower bound of the MWW. The definition of the tectonic factor is:

$$t_f = \frac{S_H - s_h}{OBG - S_h} \tag{6.3}$$

where S_H is the maximum horizontal stress. When $t_f = 0$, $S_H = s_h$; when $t_f = 1$, $S_H = OBG$. The value of t_f is usually set between 0 and 1. In the conventional 1D analysis, the value of t_f is determined by the method of 'phenomena fitting.' The drilling report and image log of the offset well in the area of the target well are required to obtain a reasonable value of t_f with the conventional 1D method. If any breakout was found in the image logging data of the wellbore, the value of t_f will be adjusted to enable the shear failure to occur at that position. The process of determining t_f is then a rather experience-dominated process. In practice, specific geostructures significantly influence the value of t_f in the region. However, because of its 1D property limitations, the conventional 1D method usually cannot include geostructural factors in the value of t_f. Conversely, the 3D FEM can build geostructures into the model; consequently, the geostructures influence the SFG calculation.

6.2.2 *Numerical procedure for calculating MWW with 3D FEM*

The existing literature does not provide a standard procedure for the calculation of MWW with 3D FEM. In practice, the following numerical procedure was used by Shen (2009) and by Shen *et al.* (2011), and is used in this calculation:

- Step 1: Build the global model at the field scale with geological and seismic data information. The material properties and geostructure parameters are defined in this step. The 3D finite element stress distribution and displacements solutions will be obtained at the field scale.
- Step 2: Using the numerical results of stress and displacement from the global model as boundary conditions, build a series of local submodels at critical TVDs, and perform numerical analyses on the submodels. (The numerical scheme for the SFG is described in Chapter 4.) For an accurate MWW solution, a primary submodel with the scale of 100 m is helpful.
- Step 3: Collate the numerical solutions of the MWW at the critical TVDs, and draw the two lines that represent SFG and FG. This is the MWW for the given trajectory.

The greater the number of submodels used in the calculation, the greater the accuracy of the MWW solution.

In the following example, seven submodels were used, which is the minimum that should be used for illustration purposes. In practice, more than 10 submodels should be used. The positions of the submodels should be carefully selected at critical positions, such as the interface between different formations or at the turning point along the trajectory where significant mud weight changes could occur.

6.3 GLOBAL MODEL DESCRIPTION AND NUMERICAL RESULTS

6.3.1 *Model description*

Figure 6.1 shows a sectional view of the field to be investigated. The pay zone is at the salt base formation. The trajectory of the wellbore penetrates the salt body and enters the pay zone at a high angle.

Figure 6.2 shows the field scale model as a block with a height of 9000 m, width of 8000 m, and length of 8,000 m. The boundary conditions of the global model include the following: all four lateral displacement constraints are applied, along with zero displacement constraints to its bottom. The top surface is free. The gravity load balanced with the initial geostress field is applied to every element of the model.

Figure 6.3 shows the salt body in red. The formation thickness below the salt to the base of the model ranges from 2200 to 2850 m. To simplify the example without losing accuracy, the

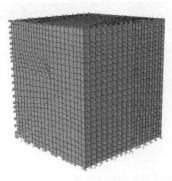

Figure 6.2.   Global model at the field scale: geometry and boundary conditions.

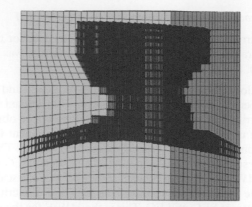

Figure 6.3.   Relative position of salt body to the formations in the model.

Figure 6.4.   Geometry of the salt body in the model.

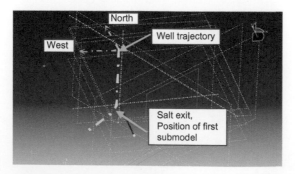

Figure 6.5.   Wellbore trajectory and the direction of salt's central axis along with salt exit of the wellbore.

formation details below the TVD interval of the reservoir are neglected. The salt geometry is a key factor that controls the stress patterns in its area. Figure 6.4 shows the salt body. Its thickness along wellbore trajectory is 5300 m. The width is 6 km, and its axis is in the direction of N30°W, as shown in Figure 6.5 and Figure 6.1.

The salt geometry adopted in the model is only a part of the actual salt body, which is far larger than that of the model. The reason for selecting only a portion of the salt body is that the other portion does not influence the stress pattern within the formation investigated. Consequently, it is necessary to have the current simplified model to reduce the computational burden. The salt geometry of the model was also selected with reference to the geometry of the trajectory: the entire area penetrated by the trajectory has been included in the model. This is necessary and essential for calculating the submodeling; the submodels require the stress and displacement solutions from the global model as their boundary conditions.

The details of the trajectory geometry are omitted from the global model for simplification without accuracy loss of the field scale modeling.

It is important to include the geostructure characteristics, such as syncline or anticline, in the model; these geostructural traits determine the stress pattern at the salt base formation. If all other conditions are fixed, the variation of the geostructure from syncline to anticline will cause obvious different stress pattern distributions at the salt base formation.

As shown in Figure 6.6, the model consists of four kinds of materials: top layer, surrounding rock, salt, and base formation. The reservoir is a part of the base formation. Table 6.1 lists the values of the material parameters.

Pore pressure data was generated with the user subroutine. Figure 6.7 shows the variation of pore pressure with TVD.

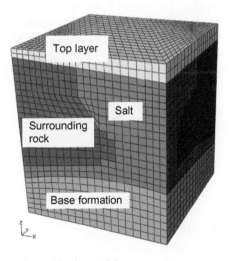

Figure 6.6.　Various materials used in the model.

Table 6.1.　Values of material parameters.

Materials	Density kg/m³	Young's modulus Pa	Poisson's ratio	CS/Pa	Friction angle
Top layer	1,900	1×10^{10}	0.3	elastic	
Salt	2,250	1.3×10^{10}	0.22	4×10^6	20°
Surrounding rock	2,350	Depth dependent	Depth dependent	1×10^6	25°
Base formation	2,350	Depth dependent	Depth dependent	4×10^6	25°

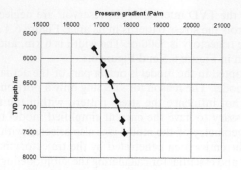

Figure 6.7. Distribution of pore pressure with TVD depth.

The code of the subroutines used in the calculation is listed below:

```
      SUBROUTINE UVARM(UVAR,DIRECT,T,TIME,DTIME,CMNAME,ORNAME,
     1 NUVARM,NOEL,NPT,LAYER,KSPT,KSTEP,KINC,NDI,NSHR,COORD,
     2 JMAC,JMATYP,MATLAYO,LACCFLA)
      INCLUDE 'ABA_PARAM.INC'
C FUNCTION OF THIS SUBROUTINE IS TO CALCULATE THE TOTAL STRESS
VALUE
      CHARACTER*80 CMNAME,ORNAME
      CHARACTER*3 FLGRAY(15)
      DIMENSION UVAR(NUVARM),DIRECT(3,3),T(3,3),TIME(2)
      DIMENSION ARRAY(15),JARRAY(15),JMAC(*),JMATYP(*),COORD(*)
C    THE DIMENSIONS OF THE VARIABLES FLGRAY, ARRAY AND JARRAY
C    MUST BE SET EQUAL TO OR GREATER THAN 15.
       CALL GETVRM('S',ARRAY,JARRAY,FLGRAY,JRCD,JMAC,JMATYP,
     1 MATLAYO,LACCFLA)
      UVAR(1) = ARRAY(1)
      UVAR(2) = ARRAY(2)
      UVAR(3) = ARRAY(3)
      UVAR(4) = ARRAY(4)
      UVAR(5) = ARRAY(5)
      UVAR(6) = ARRAY(6)
      UVAR(7) = ARRAY(1)/ARRAY(2)
       CALL GETVRM('POR',ARRAY,JARRAY,FLGRAY,JRCD,JMAC,JMATYP,
     1 MATLAYO,LACCFLA)
      UVAR(1) = -ARRAY(1)+UVAR(2)
      RETURN
      END

      SUBROUTINE USDFLD(FIELD,STATEV,PNEWDT,DIRECT,T,CELENT,
     1 TIME,DTIME,CMNAME,ORNAME,NFIELD,NSTATV,NOEL,NPT,LAYER,
     2 KSPT,KSTEP,KINC,NDI,NSHR,COORD,JMAC,JMATYP,MATLAYO,LACCFLA)
C
      INCLUDE 'ABA_PARAM.INC'
C FUNCTION OF THIS SUBROUTINE IS TO INTRODUCE PRESSURE/DEPTH
DEPENDENT PROPERTY OF E
C AND N.
      CHARACTER*80 CMNAME,ORNAME
      CHARACTER*3 FLGRAY(15)
      DIMENSION FIELD(NFIELD),STATEV(NSTATV),DIRECT(3,3),
     1 T(3,3),TIME(2),STRESS(6)
```

```
      DIMENSION ARRAY(15),JARRAY(15),JMAC(*),JMATYP(*),COORD(*)
C
      CALL SINV(STRESS,SINV1,SINV2,NDI,NSHR)
      FIELD(1) = SINV1/3.D0
C
      STATEV(1) = FIELD(1)
C
   RETURN
   END

C
   SUBROUTINE DISP(U,KSTEP,KINC,TIME,NODE,NOEL,JDOF,COORDS)
C
C FUNCTION: TO ASSIGN INITAL PORE PRESSURE IN FORMATION.
C
      INCLUDE 'ABA_PARAM.INC'
C
      DIMENSION U(3),TIME(2),COORDS(3)
C
      ZC = COORDS(3)
C
      IF(ZC.LE.2866.0.AND.ZC.GT.2190.0)
     *  U(1) = 113.73E6+(2866-ZC)/(2866.0-2190.0)*(132.26E6-
            113.73E6)
      IF(ZC.LE.2190.0.AND.ZC.GT.1407.0)
     *  U(1) = 132.26E6+(2190-ZC)/(2190.0-1407.0)*(147.22E6-
            132.26E6)
      IF(ZC.LE.1407.0.AND.ZC.GT.889.0)
     *  U(1) = 147.22E6+(1407-ZC)/(1407-889)*(158.5E6-147.22E6)
      IF(ZC.LE.889.0) U(1) = 158.5E6
      RETURN
      END
```

The depth unit used in the user subroutines **DISP** and **UPOREP** is m. Its coordinate is the global Cartesian coordinate, which has the z-axis upward, rather than the downward coordinate usually used in the petroleum industry. This is used to match the convention used in Abaqus, which is a version of solid mechanics.

```
c
      SUBROUTINE UPOREP(UW0,COORDS,NODE)
c
      INCLUDE 'ABA_PARAM.INC'
c
      DIMENSION COORDS(3)
C FUNCTION OF THIS SUBROUTINE IS TO INTRODUCE BOUNDARY CONDITION
OF PORE PRESSURE
      ZC = COORDS(3)
C
      IF(ZC.LE.2866.0.AND.ZC.GT.2190.0)
     *  UW0 = 113.73E6+(2866-ZC)/(2866.0-2190.0)*(132.26E6-
            113.73E6)
      IF(ZC.LE.2190.0.AND.ZC.GT.1407.0)
     *  UW0 = 132.26E6+(2190-ZC)/(2190.0-1407.0)*(147.22E6-
            132.26E6)
      IF(ZC.LE.1407.0.AND.ZC.GT.889.0)
```

```
*   UW0 = 147.22E6+(1407-ZC)/(1407-889)*(158.5E6-147.22E6)
    IF(ZC.LE.889.0) UW0 = 158.5E6
    RETURN
    END
```

The depth-dependent Young's modulus and Poisson's ratio are actually addressed in the model as mean-stress-dependent. The following data lines used in the model provide the model details. Accordingly, the following sentences must be used to introduce the user subroutines into the calculation.

```
**
** MATERIALS
**
*Material, name=FORMATION
*Density
2200.,
*Depvar
    1,
*Elastic, dependencies=1
1.5e11,    0.3, -1.e9
1.5e10,    0.12, -1000.
*User Defined Field
*Permeability, specific=0.1
1e-10, 0.1
1e-06, 0.5
*Drucker Prager
44.,    1., 30.
*Drucker Prager Creep, law=TIME
2.5e-22, 2.942, -0.2
*Drucker Prager Hardening, type=SHEAR
1.56e+06,0.
*Material, name=ROCK
*Density
2200.,
*Depvar
    1,
*Elastic, dependencies=1
1.5e11,    0.3, -1.e9
1.5e10,    0.12, -1000.
*User Defined Field
*USER OUTPUT VARIABLES
8
*Mohr Coulomb
25.,18.
*Mohr Coulomb Hardening
1000000., 0.
1000010., 1.0
*Material, name=SALT
*Density
2170.,
*Elastic
1.3e+10, 0.22
*Drucker Prager
44., 1., 30.
```

```
*Drucker Prager Creep, law=TIME
2.5e-22, 2.942, -0.2
*Drucker Prager Hardening, type=SHEAR
4e+06,0.
*Material, name=TOP-ROCK
*Density
1780.,
*Elastic
1e+10, 0.3
*Mohr Coulomb
25.,18.
*Mohr Coulomb Hardening
1000000., 0.
1000010., 1.0
**
*Initial Conditions, type=Stress, GEOSTATIC
Rock-2.top-layer,-1.18e7,8777.,-1.6396e7,8135.,0.8,0.8
Rock-2.rock,-1.7973e7,8555.,-1.29e8,2504.,0.8,0.8
Rock-2.salt,-1.6e7,8500.,-1.46e8,2003., 0.99,0.99
Rock-2.formation,-0.15e8,2850.,-.325e8,0.,0.8,0.8
****Rock-2.formation,-1.2883e8,2850.,-2.0323e8,0.,0.85,0.85
*Initial Conditions, type=Ratio
Rock-2.formation, 0.3
****Initial Conditions, type=Pore pressure
*Initial Conditions, type=Pore pressure, user
**
** STEP: Step-1
**
*Step, name=Step-1, nlgeom=YES
*Geostatic
**
** BOUNDARY CONDITIONS
**
*Boundary
_PickedSet14, 1, 1
_PickedSet18, 2, 2
_PickedSet16, 3, 3
*Boundary, user
Rock-2.FormationPlus,8,8
**
*model change, remove
RemoveTop
***
** LOADS
**
** Name: Gravitiy   Type: Gravity
*Dload
_PickedSet19, GRAV, 9.8, 0., 0., -1.
** Name: TopPressure   Type: Pressure
*Dsload
Surf-1, HP, 1.2e+07, 9777., 8557.
**
** OUTPUT REQUESTS
```

```
**
*Restart, write, frequency=1
**
** FIELD OUTPUT: F-Output-1
**
*Output, field, variable=PRESELECT
**
** HISTORY OUTPUT: H-Output-1
**
*Output, history, variable=PRESELECT
*End Step
**
** STEP: Step-2
**
*Step, name=Step-2, nlgeom=YES, amplitude=RAMP
*Soils, utol=2e+01
0.01, 1.0, 1.e-3
**
*Boundary, user
Rock-2.FormationPlus,8,8
**
** OUTPUT REQUESTS
**
*Restart, write, frequency=0
**
** FIELD OUTPUT: F-Output-1
**
*Output, field, variable=PRESELECT
*Element Output, directions=YES
FV, MFR, SDV, STATUS, STATUSXFEM, UVARM
**
** HISTORY OUTPUT: H-Output-1
**
*Output, history, variable=PRESELECT
*End Step
```

For the salt formation, this analysis uses the modified Drucker-Prager yielding criterion. Cohesive strength and frictional angle of the Drucker-Prager model are provided by the following values: $d = 4$ MPa, $\beta = 30°$.

The creep law, given in the following equation (Dassault Systems 2008), is adopted by:

$$\dot{\bar{\varepsilon}}^{cr} = A\left(\bar{\sigma}^{cr}\right)^{n} t^{m} \tag{6.4}$$

where $\dot{\bar{\varepsilon}}^{cr}$ represents the equivalent creep strain rate; $\bar{\sigma}^{cr}$ represents the von Mises equivalent stress; t is total time variable; and A, n, and m are three model parameters, which are given the following values:

$$A = 2.5 \times 10^{-22}, n = 2.942, m = -0.2$$

6.3.2 *Numerical results of the global model*

Figure 6.8 through Figure 6.10 show the numerical results obtained with the global model. Figure 6.8 shows the distribution of stress σ_x. Figure 6.9 and Figure 6.10 provide the distribution of stress σ_y and σ_z. The plane in the sectional view is chosen at the place where

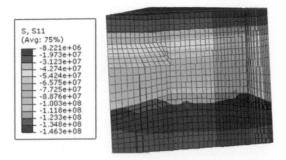

Figure 6.8.   Distribution of stress σ_x in Pa, sectional view.

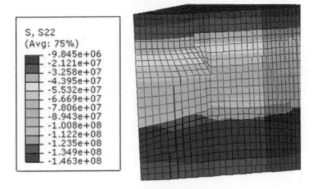

Figure 6.9.   Distribution of stress σ_y in Pa, sectional view.

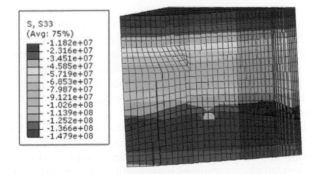

Figure 6.10.   Distribution of stress σ_z in Pa, sectional view.

wellbore trajectory is included. The stress shown here is the effective stress in which the amount of pore pressure is not included. The sign convention of solid mechanics is followed, which is positive for tensile stress and negative for compression.

6.4   SUBMODEL DESCRIPTION AND NUMERICAL RESULTS

6.4.1   *Model description*

To overcome the scale discrepancy between the field scale and wellbore section scale, two types of submodels were adopted. As shown in Figure 6.11, the size of the primary submodel

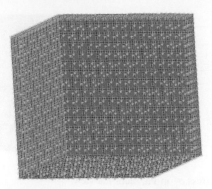

Figure 6.11.　The primary submodel.

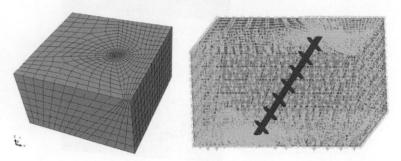

Figure 6.12.　The secondary submodel (left); the wellbore will be drilled in the SFG calculation (right).

has a width and length of 100 m and thickness is 110 m. As shown in Figure 6.12, the size of the secondary submodel has a width and length of 15 m and a thickness of 10 m. To have a reasonable MWW curve, it is necessary to have at least six sets of submodels. Because the wellbore trajectory does not change its incline angle with depth, the six sets of submodels will have the same geometry, but different specific coordinates for positions at each depth interval simulated. For brevity, the details of the submodel description are provided for only one submodel set used in this chapter.

The function of the primary submodel is only to transfer the initial geostress field and displacement boundary conditions obtained with the global model to the secondary submodel. Thus, it is not necessary for the primary submodel to include a wellbore.

The function of the secondary submodel is more important than that of the primary submodel; it is used to calculate both the SFG and FG values of the given wellbore. Consequently, the wellbore must be included in the geometry of the secondary submodels. The FG is the value of the minimum horizontal principal stress obtained with secondary submodel before the drilling of the wellbore (Figure 6.12); the SFG is the minimum value of mud weight gradient that can ensure an elastic drilling process.

In the calculation of the SFG value for the given wellbore, the drilling process was simulated in the calculation to determine the stress variation caused by the drilling process. Consequently, the model must include the actual wellbore geometry (size, shape, and direction).

The FG required in the MWW is actually selected as the minimum horizontal stress of the initial geostress field. For an inclined wellbore, it is the minimum amount of compressive stress in the petroleum industry. Consequently, the calculation of the FG requires that the wellbore not be drilled.

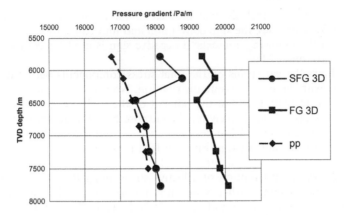

Figure 6.13. Numerical results of the SFG and FG values obtained with the secondary submodel.

In this chapter, the drilling of the wellbore will be simulated for the calculation of the SFG value, but it will not be simulated in the calculation of the FG value. Chapter 4 includes details about the calculation of the SFG.

6.4.2 *Numerical results of SFG and FG obtained with the secondary submodel*

In this subsection, the convention of the petroleum industry is adopted for the units of the variables, with 1 ft = 0.3048 m, and 1 ppg = 1174.3 Pa/m. As shown in Figure 6.13, the light line with smaller dots represents the pore pressure distribution in the reservoir formation. The location at which the wellbore trajectory exits the salt body is at 5791.2 m (19,000 ft) TVD. The calculation of the MWW focuses on the interval sections below the salt exit. Neglecting the details of the intermediate process, the numerical results for the SFG and FG values obtained with the submodel previously described, along the given wellbore trajectory, are shown in Figure 6.13. The thick, solid curve with larger dots is the SFG, which represents the lower bound of the MWW; the thinner solid curve with square marks is the FG, which represents the upper bound of the MWW.

Figure 6.13 shows that the MWW at the central region of the salt base formation is significantly larger than the values beyond the right center of the salt base formation. This indicates that the stress pattern in the central region of salt base is specific, and this aspect requires further investigation. Because the value of the effective stress ratio is also an important data input for 1D analytical tools, such as Drillworks, the effective stress ratio for the salt base region will be analyzed in detail in the following sections.

6.5 STRESS PATTERN ANALYSIS FOR SALTBASE FORMATION

The goal of this section is to perform an analysis on the stress pattern in terms of the numerical results obtained with the global model by using FEM. As a measure of the stress pattern, the effective stress ratio between three normal stress components of a stress tensor will be visualized and analyzed for the subsalt section of a wellbore. The effective stress ratio is one of the essential parameters used with 1D analytical tools for MWW design, such as Drillworks. Therefore, the stress pattern analysis in terms of the effective stress ratio will provide further understanding and useful information that can be referenced by the 1D MWW design. Focus has been put on the stress pattern in the sectional area where the wellbore trajectory was included. The red vertical line shown in Figure 6.14 is selected as Path-1, along which the stress distribution will be plotted.

For the convenience of analysis, which is based on mathematical description in the framework of solid mechanics, the sign convention and unit convention of solid mechanics will be adopted in the following contexts: a negative sign is used for compression and the stress is shown in the unit of Pa.

Figure 6.15 shows the variations of the total stress components with z-depth from the top surface along with the pore pressure distribution. Non-zero pore pressure occurs at the top of the reservoir formation and at the salt exit of the wellbore trajectory. The effective stress component, σ_z, is provided solely for the purpose of comparison. A close-up view of Figure 6.15, which focuses on a 2-km depth interval around the salt-formation interface is shown in Figure 6.16. As illustrated in Figure 6.16, the vertical stress component, σ_z, is the minimum stress component in this depth interval.

Figure 6.17 shows the distribution of effective stress along Path-1. The distribution of the effective stress ratios along Path-1 is shown in Figure 6.18. A close-up view around the salt-reservoir interface for the curves in Figure 6.18 is shown in Figure 6.19. For points in the salt base along Path-1, the effective stress ratios are significantly greater than 1. This result, however, is only true for the Path-1. For a different path, the effective stress ratio will be different. As shown in Figure 6.19, there is a range of 1.5 km within which the effective stress ratio is greater than 1. Because the absolute values of stress components are in the sequence of $\sigma_x > \sigma_y > \sigma_z$, its stress pattern is the reverse faulting pattern, even though no fault structure exists there.

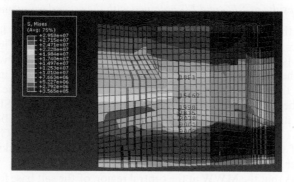

Figure 6.14.   Path-1 along which the stress distribution will be plotted.

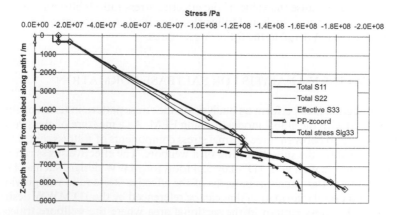

Figure 6.15.   Variation of stress components with z-depth from the top surface.

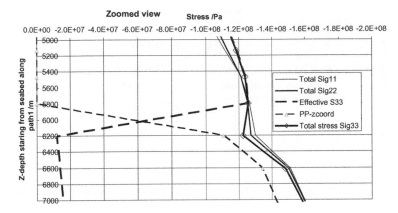

Figure 6.16.   Zoomed-in view of stress variation with depth.

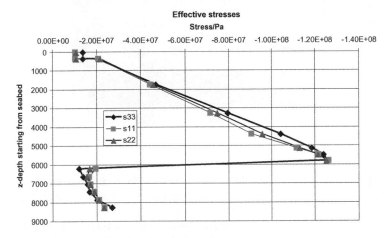

Figure 6.17.   Distribution of effective stress ratio along Path-1.

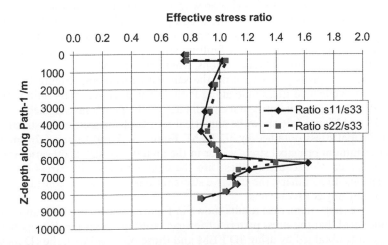

Figure 6.18.   Distribution of effective stress ratio along Path-1.

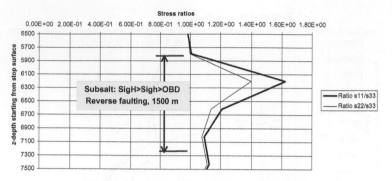

Figure 6.19.   Distribution of effective stress ratio along Path-1: Zoomed-in view around the salt-reservoir interface.

Figure 6.20.   Illustration of Path-2.

To better explain the range of area in which the reverse faulting stress pattern exists within the subsalt formation, investigations of the distribution of stress components along Path-2 and Path-3 are presented in the following contexts.

As shown in Figure 6.20, Path-2 is selected within the salt body. Figure 6.21 shows that the stress distribution along Path-2 inside the salt body is in the order of $\sigma_z > \sigma_y > \sigma_x$, which indicates a normal faulting stress pattern.

As shown in Figure 6.22, Path-3 is chosen within the subsalt formation and is horizontal. Figure 6.23 shows the distribution of stress components along Path-3. Figure 6.23 illustrates a region (marked with a blue-dashed circle) whose stress components is in the order of $\sigma_x > \sigma_y > \sigma_z$, which represents the reverse faulting stress pattern. Other locations outside the reverse faulting circle are normal fault regions, where $\sigma_z > (\sigma_y$ and $\sigma_x)$.

To investigate the stress pattern within the salt base formation, Figure 6.24 shows the distribution of the effective stress ratio along Path-3. The effective stress ratio varies significantly from 0.65 to 1.23 with horizontal coordinates of the points investigated.

Figure 6.25 and Figure 6.26 show the numerical results of the sectional distribution of the minimum principal stress obtained by 3D FEM. Figure 6.25 and Figure 6.26 show that the direction of minimum principal stress vector in the area of the salt body varies significantly from place to place. Consequently, these are the causes of effective stress ratio variation within formations. In summary, the effective stress ratio is by far not a constant, and cannot be represented by any constant value.

With reference to the numerical results shown in Figure 6.19, Figure 6.23, and Figure 6.24, it can be concluded that the stress pattern of reverse faulting causes the discrepancy between the MWW results obtained by using 3D FEM and those obtained with the 1D method.

Because no faulting structure exists in the subsalt formation, it is difficult to relate its stress pattern to the form of reverse faulting. Consequently, in the input data of the 1D calculation

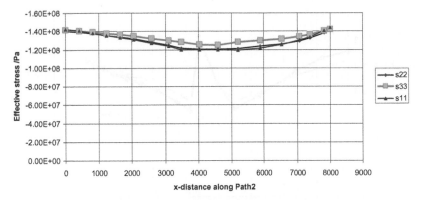

Figure 6.21.   Distribution of effective stress components along Path-2.

Figure 6.22.   Illustration of Path-3.

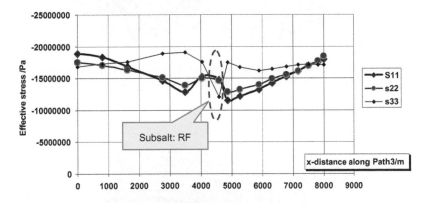

Figure 6.23.   Distribution of stress components along Path-3.

for the **MWW**, the effective stress ratio can usually be classified as normal faulting. However, this kind of stress pattern can be discovered using 3D FEM numerical calculations, although it is not used as input data there.

Figure 6.27 shows an example of mud weight logging data reported by Shen (2009). The mud weight used in the drilling of the subsalt wellbore section (the black curve) is greater than the overburden gradient. This indicates that the minimum horizontal stress component is greater than the vertical component and, consequently, provides an example of a reverse faulting stress pattern in the subsalt formation.

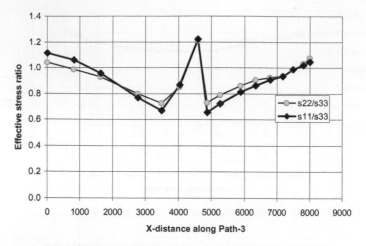

Figure 6.24.   Effective stress ratio along horizontal Path-3.

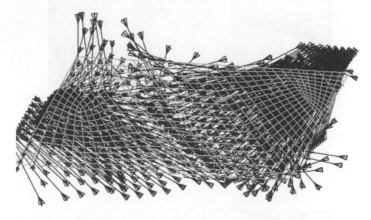

Figure 6.25.   Finite element results: the sectional view of the minimum principal stress at the salt base formation with TVD = 6142 m (z-coordinate = 2858 m) in 3D space.

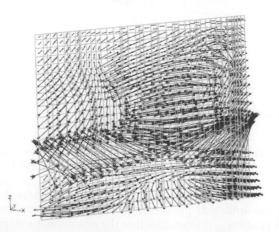

Figure 6.26.   Finite element results: the sectional view of the minimum principal stress in the plane which is normal to the central axis of salt body in 3D space.

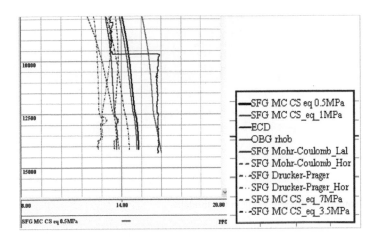

Figure 6.27.   An example of mud-weight logging data used in practice.

6.6   ALTERNATIVE VALIDATION ON STRESS PATTERN WITHIN SALTBASE FORMATION

Because the distribution of the effective ratio within the saltbase formation differs from the expected scenario, an alternative stress pattern validation was performed. The alternative validation included altering the salt geometry. Although the geometry of a salt body is usually determined by a seismic expert, it is possible to make a different decision at this aspect. The salt geometry with the alteration is shown in Fig. 6.28. In comparison to its original form shown in in Fig. 6.3 and Fig. 6.4, the bottom of the salt body has been changed.

The numerical results of the effective stress ratios obtained with the altered salt geometry are shown in Fig. 6.29 and Fig. 6.30 respectively. The values of stress ratios range from 1.4 to 0.4. The patterns of distribution of the effective stress ratios are similar to that obtained with the previous salt geometry.

This result indicates that the influence of the anticline geostructure on the stress pattern is significant, and should not be ignored in the MWW design.

6.7   A SOLUTION WITH 1D TOOL DRILLWORKS AND ITS COMPARISON WITH 3D SOLUTION

The results of SFG and FG values obtained with 1D software are shown in Fig. 6.31 for comparison purposes. The dashed, dark blue curve is the SFG solution obtained using the 1D method, and the solid blue curve is the FG solution obtained with the 1D method.

The input data used for the 1D solution includes the following:

- Effective stress ratio: the value of k0 = 0.8 is used for the entire well section.
- Tectonic factor: the value of $t_f = 0.5$ is used for the entire well section.
- Overburden gradient is given by the same density used in 3D model and is equivalent to the vertical stress of the 3D model at the TVD of the salt exit; the strength parameters are the same as those used by the 3D model.
- The pore pressure shown in Figure 6.7 is adopted in the calculation.

A comparison between the 1D solution with its corresponding 3D MWW solutions shows that a significant difference exists between the two in the TVD interval from 5,000 to 8,000 m. The two sets of MWW results are almost the same beyond the previously described TVD. The maximum difference between the two sets of SFG values is approximately 1,300 Pa/m,

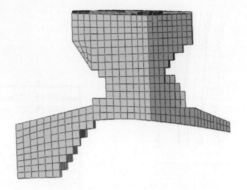

Figure 6.28.   Geometry of the salt body in the model.

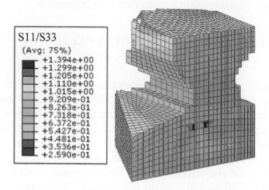

Figure 6.29.   Effective stress ratio distribution of σ_x/σ_z.

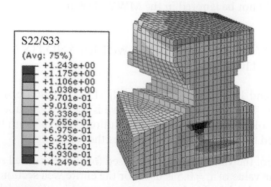

Figure 6.30.   Effective stress ratio distribution of σ_y/σ_z.

which occurs at the TVD of approximately 6,130 m. Accounting for the depth value, the mud weight pressure difference obtained by the 3D and 1D methods are significant, and the discrepancy between the value of the effective stress ratio used in the 1D analysis and that implicitly used in the 3D analysis is believed to be the major reason for this phenomena.

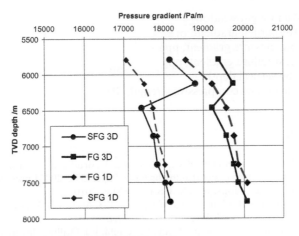

Figure 6.31. Comparison between 1D and 3D MWW solutions.

6.8 CONCLUSIONS

Using FEM, a 3D numerical solution of the MWW was obtained for a subsalt wellbore section. Based on the numerical results presented in this chapter, the following conclusions can be obtained:

- There is a region of the subsalt formation where the reverse faulting stress pattern may be formed. It is most likely to be true, particularly when an anticline structure exists at the salt base.
- As the wellbore trajectory penetrates this region, the FG and SFG values at this region may appear as abnormally high values. Consequently, the MWW obtained for this wellbore trajectory shown in the results is shifted/greater than the normal values as compared to the 1D solution obtained with 1D software.
- The numerical results indicate that the effective stress ratio within the salt base formation changes with TVD and with the horizontal position. For an accurate MWW solution for subsalt wellbore sections, the 3D FEM tool should be used in the design of the MWW.

NOMENCLATURE

A	=	Model parameter in creep model
n	=	Model parameter in creep model
m	=	Model parameter in creep model
k_0	=	Effective stress ratio
S_h	=	Minimum horizontal stress component, Pa
S_H	=	Maximum horizontal stress component, Pa
t_f	=	Tectonic factor
σ_x	=	Effective stress component in x-direction
σ_y	=	Effective stress component in y-direction
σ_z	=	Effective stress component in z-direction
ρ	=	Density, kg/m^3
CS	=	Cohesive strength, Pa
FA	=	Internal friction angle, °
FEM	=	Finite Element method
FG	=	Fracture gradient, ppg

MWW = Mud weight window
OBG   = Overburden gradient, ppg
pp   = Pore pressure gradient, ppg
SFG   = Shear failure gradient, ppg
TVD   = True vertical depth, m
UCS   = Unconfined compressive strength, Pa

REFERENCES

Cullen, P.C., Taylor, J.M.R., Thomas, W.C., Whitehead, P., Brudy, M. and vander Zee, W.: Technologies to identify salt-related deep-water drilling hazards. Paper OTC 20854 presented at the 2010 Offshore Technology Conference, Houston, TX, USA, 3–6 May, 2010.

Dassault Systems: *Abaqus Analysis User's Manual.* Vol. 3: Materials, Version 6.8, Vélizy-Villacoublay, France: Dassault Systems, 19.3.1-17 – 19.3.2-14, 2008.

Dusseault, M.B., Maury, V., Sanfilippo, F. and Santarelli, F.J: Drilling through salt: constitutive behavior and drilling strategies. Paper ARMA/NARMS 04–608 Gulf Rock 2004 presented at the 6th North America Rock Mechanics Symposium, Houston, TX, 5–9 June, 2004.

Fredrich, J.T., Coblentz, D., Fossum, A.F. and Thorne, B.J.: Stress perturbation adjacent to salt bodies in the deepwater Gulf of México. Paper SPE 84554 presented at the Annual Technical Conference and Exhibition, Denver, CO, USA, 5–8 October, 2003.

Power, D., Ivan, C.D. and Brooks, S.W.: The top 10 lost circulation concerns in deepwater drilling. Paper SPE 81133 presented at the SPE Latin American and Caribbean Petroleum Engineering Conference, Port of Spain, Trinidad, West Indies, 27–30 April, 2003.

Shen, X.P.: DEA-161 *Joint Industry Project to Develop an Improved Methodology for Wellbore Stability Prediction: Deepwater Gulf of Mexico Viosca Knoll 989 Field Area.* Halliburton Consulting, Houston, TX, USA, 18 August, 2009.

Shen, X.P., Diaz, A. and Sheehy, T.: A case study on mud-weight design with finite-element method for subsalt sells. Paper 1120101214087 accepted by and to be presented at the 2011 ICCES, Nanjing, China, 18–21 April, 2011.

Tsai, F.C., O'Rouke, J.E. and Silva, W.: Basement rock faulting as a primary mechanism for initiating major salt deformation features. Paper 87–0621 presented at the 28th US Symposium on Rock Mechanics, Tucson, AZ, USA, 29 June-1 July, 1987.

Whitson, C.D. and McFadyen, M.K.: Lesson learned in the planning and drilling of deep, subsalt wells in the deepwater Gulf of Mexico. Paper SPE 71363 presented at the SPE Annual Technical Conference and Exhibition, New Orleans, LA, USA, 30 September-3 October, 2001.

Willson, S.M. and, Fredrich, J.T.: Geomechanics considerations for through- and near-salt well design. Paper SPE 95621 presented at the 2005 SPE Annual Technical Conference and Exhibition, Dallas, TX, 9–12 October, 2005.

CHAPTER 7

Numerical calculation of stress rotation caused by salt creep and pore pressure depletion

Xinpu Shen, Arturo Diaz & Mao Bai

7.1 INTRODUCTION

The orientation of the principal horizontal stress has an important influence on completion design (i.e., casing direction and hydraulic fracturing). Several tectonic and depositional mechanisms influence the orientation of the principal stresses:

- Relaxation of stresses adjacent to faults.
- Accumulation of stresses adjacent to faults before slippage.
- Halo kinetics (i.e., movement of salt masses).
- Slumping.
- Rapid deposition of sediments on top of a subsurface environment dominated by strike-slip or reverse faulting.

Stress rotation has been reported by several authors in various drilling environments that share common complex geological structures, including active tectonic regions, complex fault and joint systems, salt bodies, and depleted reservoirs. Stress rotation can be observed within one well or from one well to another well. It causes extremely expensive and difficult wellbore stability problems during drilling, completion, or production, and represents a challenge for the oil industry in both operation and modeling. Consequently, a great deal of effort has been expended to study and fully understand this phenomenon.

Martin and Chandler (1993) reported a maximum horizontal stress rotation near two major thrust faults that was intersected during the excavation of the Underground Research Laboratory (URL) in the Canadian Shield. In this region, the fault system divides the rock mass into varying stress domains. Above the fault system, the rock mass contains regular joint sets in which the maximum horizontal stress is oriented parallel to the major sub-vertical joint set. Below the fault system, the rock is massive with no jointing; the maximum horizontal stress has rotated approximately 90° and is aligned with the dip direction of fracture zone. Stress rotation is commonly observed where the block of rock above the fault has lost its original load because of displacement above the fault. This occurrence results in considerably less maximum horizontal stress magnitude than the magnitudes below the fracture zone, where the maximum horizontal stress magnitude is fairly constant (Martin and Chandler 1993). The minimum horizontal stress also increases in the vicinity of faults and decreases above and below the fault zone. Teufel et al. (1984) observed stress rotation with depth at the multi-well experiment site in Rifle, Colorado, from N75°W in the upper fluvial zone (at 1501 m), to N89°W in the coastal zone (at 1980 m), to N75°W in the marine zone (at 2410 m). This observation is a result of the stress associated with large local topographical loading super-imposed on the regional stress field of the basin (Teufel et al., 1984).

Last and McLean (1996) reported a significant rotation of stresses in the tectonically-active foothills of the Cusiana field in Colombia. The combined effect of the loading and structure (faults and dips) produce shear stresses in the overburden, resulting in a rotation of principal stresses away from the vertical and horizontal directions (Last and McLean 1996). Charlez et al. (1998) discussed the six tectonic regions of Colombia oil fields that present displacement along faults and large strain to account for stress rotation arising from the force balance between the Nazca, Caribbean, and South American plates. This scenario required tighter constraints on the relative principal stress magnitude in the Cusiana field, where the principal stresses rotate near active faults to satisfy the friction condition along the fault plane.

Miskimins *et al.* (2001) observed stress rotation in a complex faulting field (North LaBarge Shallow Unit of the Green River basin), which is located in a major thrust fault, a tear fault, four major strike-slip faults, and a "horse-tail" splay termination. The regional maximum stress orientation is consistent with the north-south direction; however, the wells associated with the thrust fault and major strike-slip exhibit stress rotation between 45° and 90° as a result of the proximity to the major strike-slip and reverse fault, which is likely attributable to changes in tectonic stresses through geological time.

Depletion can also induce stress rotation and impose high pressure on casing by direct compaction of the reservoir rock, overburden fault, and bedding plane movement triggered by the reservoir compaction and in-well mechanical hot spots. Kristiansen (2004) discussed the compaction of the chalk formation in Valhall field, which caused a seafloor subsidence of nearly 5 m and stress reorientation of approximately 90°. The subsidence continues at 0.25 cm/year, and a major effort has been expended to understand the large effect of wellbore stability that creates the surface changes. The surface changes are followed by depletion, compaction, and subsidence, which also reactivate faults and shear casing.

The perturbation of stresses adjacent to salt bodies is significant and depends on the geometry of the salt. Salt affects the actual geomechanical environment through the alteration of the local stresses because salt cannot sustain deviatoric stress and deforms by means of plastic creep in response to any imposed deviatoric stress. Consequently, changes in the stress field and shear stresses adjacent to the salt are sufficient to cause a reorientation of the principal stress. It is well known that rubble zones occur below or adjacent to the salt diapirs. It is an intrinsic consequence of the equilibrium between the stress field within the salt body and that within the adjacent formation. Fredrich *et al.* (2003) reported the results of a 3D finite element method (FEM) simulation of various geometries of salt bodies, including spherical salt body, horizontal salt sheet, columnar salt diapir, and columnar salt diapir with an overlaying tongue. The results show the following:

- The shear stress is highly amplified in certain zones in specific geometry adjacent to the salt.
- There is induced anisotropy of horizontal stresses.
- Orientations of principal stresses are no longer vertical and horizontal, and the vertical stress may not be the maximum stress component.
- For some geometries, vertical stress within and adjacent to the salt is not equal to the gravitational load (stress-arching effect occurs) before drilling or production and behave similarly to the overburden changes during the depletion of compactable reservoirs.
- The stress perturbation can only be determined by solving the complete set of equilibrium, compatibility, and constitutive equations with the appropriate initial and boundary condition using the FEM (Fredrich *et al.*, 2003).

Works published on applications of principal stress directions are rich, but very few works reported on the analysis of the mechanism of stress orientation, neither experimental nor analytical. Goals of this chapter include the following:

- Review the phenomena of stress rotation in the presence of faulting, salt bodies, and non-uniform distribution of physical rock properties, pore pressure, and tectonic deformation.
- Use a 3D FEM to numerically simulate the stress-field distribution for these cases at the field scale (the tectonic stress factor and gravity loads are included in the loading used in the analysis).
- Numerically analyze the variation of stress orientation within formations caused by enhancement measures, such as liquid injection.

7.2   STRESS ANALYSIS FOR A SUBSALT WELL

This section presents the results of the numerical analysis for a subsalt well, which penetrated a thick salt body.

The fundamental issue for a successful application of the numerical method in the analysis of stress orientation is the accurate entry of the pore pressure values encountered along the wellbore. A second issue affecting success is the accurate description of the 3D structure of the salt body and related formations, which are used in the calculation.

One-dimensional (1D) analytical software, such as the Drillworks® software, has proven to be a very powerful analytical tool used in the prediction of pore pressure for many wellbores around the world (Shen 2009). These kinds of tools have provided a good foundation for a successful 3D numerical analysis of stress orientation. Modern seismic technology can also provide accurate geological and structural information for the salt dome/canopy and related formations. Consequently, it is necessary and practical to perform a 3D numerical analysis on salt-body-related stress orientation surrounding a salt body.

7.2.1   *Computational model*

Fig. 7.1 shows the geometric profile of the model. Its width and thickness are both 10 km; its height on the left side is 10 km, and its height on the right side is 9.6 km, which shows a variation of ground surface.

The geometry of the salt body was built as shown in Fig. 7.2. Its outer edge diameter is 7.01 km, and the maximum thickness is 1.676 km. Its upper surface has a 30° angle with the horizon. The depth of its top edge from the ground surface is 1.219 km.

The loads sustained by the model include gravity load distributed within the model body and pore pressure distributed in the formations. Because the salt body has no porosity or permeability, the pore pressure within the salt body is assumed to be zero. The pore pressure existing in the subsalt reservoir formation is given as 47 MPa, according to the software-predicted pore pressure value.

This analysis uses the modified Drucker-Prager yielding criterion. Values of material properties are listed below and reference has been made to earlier works, specifically to Hunter *et al.,* 2009; Infante and Chenevert 1989; Maia *et al.,* 2005; Marinez *et al.,* 2008; and Aburto and Clyde 2009. The values of strength parameters for salt adopted here by the Drucker-Prager model are $d = 4$ MPa, $\beta = 44°$, which correspond to values in the Mohr-Coulomb model, as $c = 1.25$ MPa, $\phi = 25°$.

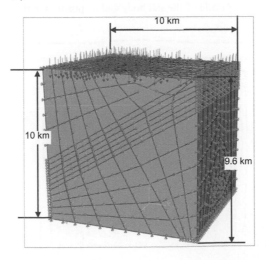

Figure 7.1.   Model geometry: profile of the entire model.

For the formation, the cohesive strength and frictional angle of Drucker-Prager model are given the following values: $d = 1.56$ MPa, $\beta = 44°$, which corresponds to values in the Mohr-Coulomb model, as $c = 0.5$ MPa, $\phi = 25°$.

The creep law, given in Eq. 7.1 (Dassault Systems 2008), is adopted as follows:

$$\dot{\bar{\varepsilon}}^{cr} = A\left(\bar{\sigma}^{cr}\right)^n t^m \tag{7.1}$$

where $\dot{\bar{\varepsilon}}^{cr}$ represents the equivalent creep strain rate; $\bar{\sigma}^{cr}$ represents the von Mises equivalent stress; t is the total time variable; A, n, m are three model parameters, which are given values of 2.5e-22, 2.942, and −0.2, respectively.

Zero-displacement boundary constraints were applied in the normal direction of the four lateral surfaces and the bottom surface.

7.2.2 *Numerical results*

Fig. 7.3 through Fig. 7.8 show the numerical results of principal stress directions of the FEM. Fig. 7.3 through Fig. 7.5 show sectional views in the Cartesian XY plane at coordinate z = 5000 m, which is the central depth of the salt body. Fig. 7.6 through Fig. 7.8 are sectional views in the Cartesian YZ plane at coordinate x = 5000 m, where it also goes through the central position of the salt body.

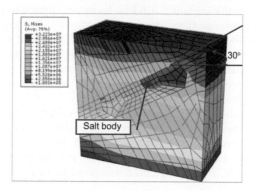

Figure 7.2.   Model geometry: profile of the salt body and its position in the model (sectional view).

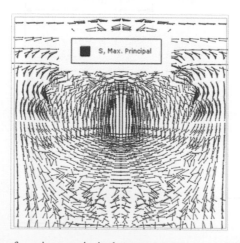

Figure 7.3.   Distribution of maximum principal stress component: z = 5000 m, Cartesian XY view.

The sign convention used in these figures follows the convention used for solid mechanics, in which tensile stress is defined as positive, rather than that used for geoengineering. Therefore, the maximum principal stress is the minimum compressive stress.

As shown in Fig. 7.3, the direction of the maximum principal stress rotates with changes of the azimuth angle in the surrounding rocks of the salt body. At a certain distance from the salt body, as shown in Fig. 7.3, the maximum principal stress becomes normal, which is straight.

Fig. 7.4 shows the direction of the medium principal stress, and it appears to be perpendicular to the direction of the maximum principal stress in this Cartesian XY view. Fig. 7.5 shows the direction of the minimum principal stress, and its vector appears to be very short because it is actually the vertical stress component.

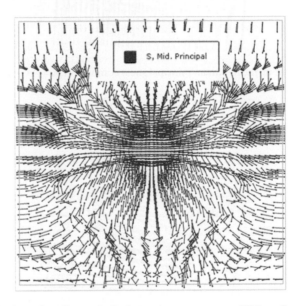

Figure 7.4.   Distribution of medium principal stress component: z = 5000 m, Cartesian XY view.

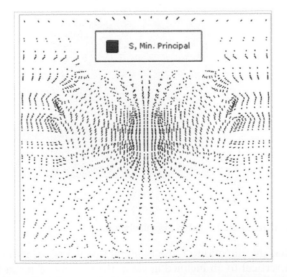

Figure 7.5.   Distribution of minimum principal stress component: z = 5000 m, Cartesian XY view.

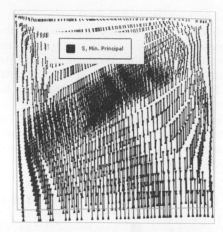

Figure 7.6.   Distribution of the minimum principal stress component: x = 5000 m, Cartesian YZ view.

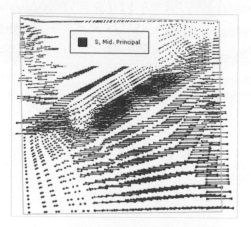

Figure 7.7.   Distribution of the medium principal stress component: x = 5000 m, Cartesian YZ view.

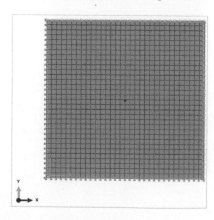

Figure 7.8.   The finite element mesh used in the calculation.

Fig. 7.6 shows that the direction of the minimum principal stress in the Cartesian YZ plane changes from vertical to an inclined angle in the vicinity of a salt body and becomes vertical at a distance away from it. In Fig. 7.7 and Fig. 7.8, the orientations of the medium and maximum principal stresses are primarily horizontal.

7.3   VARIATION OF STRESS ORIENTATION CAUSED BY INJECTION AND PRODUCTION

This section investigates the stress rotation caused by water injection and/or oil production. A two-dimensional (2D) model is used. The influence of the anisotropic property of permeability is numerically simulated and investigated. The numerical results presented in this chapter have visualized the change of stress pattern from its original isotropic stress pattern to a specific form after injection and production.

7.3.1   *The model used in the computation*

Fig. 7.8 shows the finite element mesh used in the calculation. There are a total of 3201 nodes and 1024 CPE8RP quadratic plane strain elements used in the mesh. Pore pressure and nodal displacement are two types of primary variables. The width and length are 1600 m. One well in the center of the model is represented by a dot. Injection and production are simulated with concentrated flow at the well location. A constant initial pore pressure of pp = 10 MPa was given to all of the nodes of the model, including the boundary nodes.

The following material parameters were used:

- Young's modulus $E = 1 \times 10^{10}$ Pa, Poisson's ratio $v = 0.3$
- Initial geostress is set as isotropic, i.e., $\sigma_x = \sigma_y = 10 \times 10^6$ Pa, and plane strain status is set with $\varepsilon_z = 0$
- Initial porosity $\phi = 0.25$

Normal displacement constrains were applied to the four sides of the model.

In the following paragraphs, the numerical results of principal stresses include both values and directions for four different cases, and will be given in the following sequence:

- Isotropic permeability with injection.
- Isotropic permeability with production.
- Orthotropic permeability with injection.
- Orthotropic permeability with production.

The convention used for solid mechanics signs is followed, i.e., positive for tensile stress and negative for compressive stress.

It is obvious that the direction of the initial in-plane maximum principal stress is in the horizontal direction, and the direction of the initial in-plane minimum principal stress is in the vertical direction in the 2D planar space. Thus, the visualization of their distributions is neglected here.

7.3.2   *Numerical results*

7.3.2.1   *Numerical results of stress rotation with isotropic permeability and injection*
The following results are obtained with isotropic permeability $k = 100$ Darcy. Although this number is rather high, it is physically possible; consequently, it is a reasonable value to be used in this example. The maximum injection pressure is controlled with a limit of $pp_{max} = 1.335 \times 10^7$ Pa.

Fig. 7.9 shows the distribution of pore pressure after injection. The highest pore pressure is at the center of the field. The visualization of the in-plane maximum principal stress in Fig. 7.10 shows that the direction of the maximum principal stress is in the circumferential direction, which uses the position of well as the center of the circle. Fig. 7.11 shows that the minimum principal in-plane stress is in the radial direction of the circle. These figures show the drastic change of principal directions after injection.

7.3.2.2   *Numerical results on stress rotation with isotropic permeability and production*
The following results are obtained with isotropic permeability and production. The values of parameters used here are the same as those used for injection simulation. The minimum production pressure is controlled with a limit of $pp_{min} = 6 \times 10^6$ Pa.

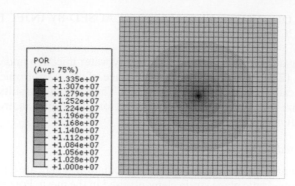

Figure 7.9.   Distribution of pore pressure after injection.

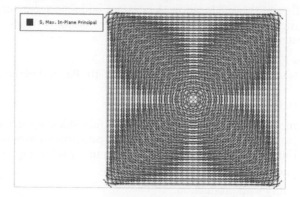

Figure 7.10.   Distribution of direction of maximum in-plane principal stress after injection.

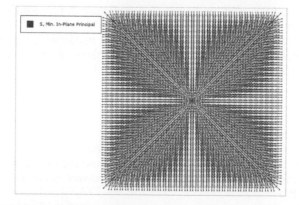

Figure 7.11.   Distribution of direction of minimum in-plane principal stress after injection.

Fig. 7.12 shows that the distribution of pore pressure after production is similar to that after injection, but with the opposite gradient of pore pressure: the lowest pore pressure is at the center of the field. Fig. 7.13 provides a visualization of the in-plane maximum principal stress and shows that the direction of the maximum principal stress is in the radial direction, which uses the position of well as the center of the circle. Fig. 7.14 shows that the minimum principal in-plane stress is in the circumferential direction of the circle. These figures demonstrate the drastic change of principal directions after injection.

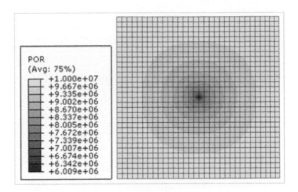

Figure 7.12.   Distribution of pore pressure after production.

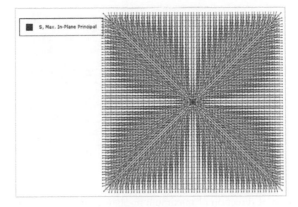

Figure 7.13.   Distribution of direction of maximum in-plane principal stress after production.

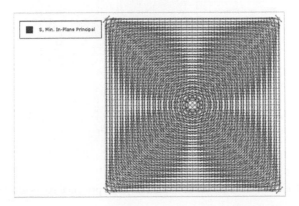

Figure 7.14.   Distribution of direction of minimum in-plane principal stress after production.

7.3.2.3   *Numerical results on stress rotation with orthotropic permeability and injection*
The following results are obtained with orthotropic permeability $k_x = 100$ and $k_y = 400$ Darcy. Although these numbers are rather large, they are physically possible. Consequently, it is are reasonable to use them in the example. The maximum injection pressure is controlled with a limit of $pp_{max} = 1.163 \times 10^7$ Pa.

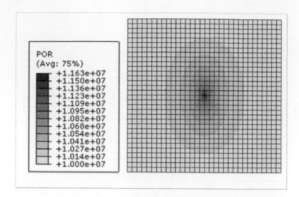

Figure 7.15.　Distribution of pore pressure after injection.

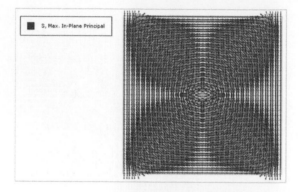

Figure 7.16.　Distribution of direction of maximum in-plane principal stress after injection.

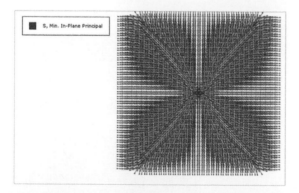

Figure 7.17.　Distribution of direction of minimum in-plane principal stress after injection.

In Fig. 7.15, the distribution of pore pressure after injection is not isotropic because it is influenced by its orthotropic property of permeability: the pore pressure increase in the y-direction occurs more quickly than the increase that occurs in the x-direction. The contour of the pore pressure distribution in the field becomes elliptical. Consequently, in the visualization of the in-plane maximum principal stress, shown in Fig. 7.16, the direction of the maximum principal stress appears to be in the circumferential direction of the ellipse, which uses the position of well as the center. Fig. 7.17 shows that the minimum principal in-plane stress is in the radial direction of the ellipse. These figures demonstrate the drastic change of

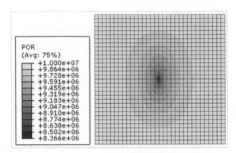

Figure 7.18. Distribution of pore pressure after production.

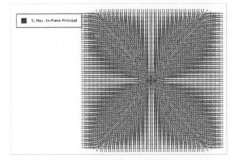

Figure 7.19. Distribution of direction of maximum in-plane principal stress after production.

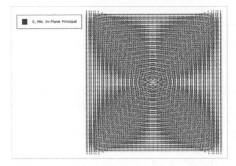

Figure 7.20. Distribution of direction of minimum in-plane principal stress after production.

principal directions after injection and the significant influence from the anisotropic property of its permeability.

7.3.2.4 *Numerical results on stress rotation with orthotropic permeability and production*

The following results are obtained with orthotropic permeability and production. The values of parameters used here are the same as those used for the orthotropic permeability and injection simulation. The minimum production pressure is controlled with a limit of $pp_{min} = 8.366 \times 10^6$ Pa.

In Fig. 7.18, the distribution of pore pressure after production is not isotropic because it is influenced by its orthotropic property of permeability: the pore pressure decrease in the y-direction occurs more quickly than the decrease that occurs in the x-direction. The contour of the pore pressure distribution in the field is elliptical. Consequently, in the visualization of the in-plane maximum principal stress, shown in Fig. 7.19, the direction of the maximum principal stress appears to be in the radial direction of the ellipse, which uses the position of well as its center. Fig. 7.20 shows that the minimum principal in-plane stress is in the circumferential direction of the ellipse. These figures illustrate the drastic change of principal directions after production and the significant influence from the anisotropic property of its permeability.

7.3.3 *Remarks*

The following conclusions can be derived from the previously described numerical results of flow-induced stress re-orientation:

- The direction of principal stress varies drastically from its initial direction during the transient injection/production process. Consequently, it is important to account for the influence of injection/production activities of nearby wells in the analysis process of wellbore stability or stimulation design.
- The anisotropy of permeability of a field will influence the effect of the injection design: pore pressure distributed in the direction that has greater permeability will be greater than that with lesser permeability values. This occurrence will cause a non-uniform distribution of pore pressure in the process of injection/production.

7.4 VARIATION OF STRESS ORIENTATION CAUSED BY PORE PRESSURE DEPLETION: CASE STUDY IN EKOFISK FIELD

This section describes the numerical model built for the Ekofisk field, the behavior of the model under geostress, and the simulation of pore pressure depletion. It also presents the results and analyses of the variation of principal stress orientation.

7.4.1 *The numerical model*

Fig. 7.21 shows the Ekofisk field. The total depth of the model is 4000 m, the width is 5500 m, and the length is 9000 m; the distribution of the chalk reservoir is shown in red. The model uses four vertical layers of overburden. The depth of the first layer is 1500 m, the second is 800 m, the third is between 435 and 800 m, and the depth of the bottom layer is between 900 and 1265 m. The reservoir layer that ranges from 50 to 150 m is located in the lower middle of Layer 3, as shown in Fig. 7.22.

As shown in Fig. 7.21, the horizontal distance between the end points of two reservoir intersections is approximately 2000 m. This distance suggests that the radial displacement from each wellbore, where the effect of pressure depletion is expected to be encountered, is approximately 1000 m. Consequently, the local pressure depletion around a wellbore is assumed to have a circular area of influence, as shown in Fig. 7.23. The horizontal distance between points A and B, shown in Fig. 9, is 2100 m; point B is located in the center of the circle area of Fig. 7.23.

The complexity of the Ekofisk chalk creates issues related to visco-plasticity (Hickman 2004) and to compatibility (Cipolla *et al.,* 2007). Furthermore, the Young's modulus values of the chalk vary with pressure in the effective stress space.

This work adopts the modified Drucker-Prager yielding criterion. The cohesive strength and frictional angle are given the following values: $c = 1$ MPa, $\phi = 25°$. The creep law, given by Eq. 7.1, is adopted.

The compaction property of the chalk reservoir is simulated with a linear law of hardening. Chalk skeleton variations of both Young's modulus and Poisson's ratio with pressure in the effective stress space are expressed in Fig. 7.24 and Fig. 7.25, respectively.

The property of pressure dependency of the chalk is determined by using the Abaqus User Subroutines in the calculation. The porosity parameters of the chalk are given the following values: initial void ratio $R = 0.5$; intrinsic permeability coefficient $k = 2$ Darcy.

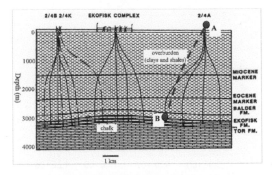

Figure 7.21. Geostructure and wellbore distribution in the area of Ekofisk (with reference to Sulak and Danielsen 1989).

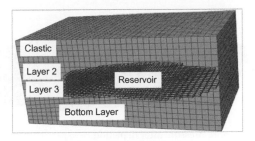

Figure 7.22. Distribution of the reservoir.

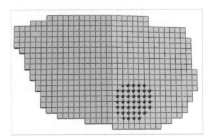

Figure 7.23. Distribution of influence area of pore pressure depletion around wellbore concerned (marked with red dots).

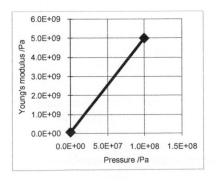

Figure 7.24. Pressure dependency of Young's modulus.

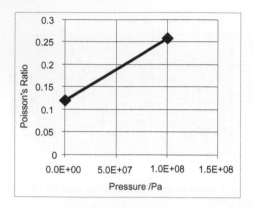

Figure 7.25.   Pressure dependency of Poisson's ratio.

Figure 7.26.   Distribution of pore pressure after depletion.

The clastic layer on the top of the model and the bottom layer material of the model are assumed to be elastic. The material of Layer 2 and Layer 3 are assumed to be visco-elasto-plastic.

The depth of overburden seawater is 100 m. The seawater produces a uniform pressure of 0.96 MPa on the overburden rock of the field-scale model. The geostress field is balanced by the gravity field in the vertical direction, and components of lateral stress are assumed to have a value of 90% of the vertical component. The density values of the reservoir and the four model layers are given as:

$$\rho_{reservoir} = 2100 \ kg/m^3, \rho_{clatic} = 2200 \ kg/m^3,$$
$$\rho_{layer-2} = 2250 \ kg/m^3, \rho_{layer-3} = 2250 \ kg/m^3, \tag{7.2}$$
$$\rho_{bottom} = 2500 \ kg/m^3$$

The initial pore pressure within the reservoir is assumed to be 34 MPa.

In this calculation, a local pore pressure depletion of 34 MPa to 10 MPa is used to simulate the subsidence caused by the production of the well studied. Fig. 7.26 illustrates the distribution of the depletion of pore pressure.

7.4.2   Numerical results

Fig. 7.27 shows the results of subsidence after pressure depletion in the area surrounding the studied wellbore. The top of the block is at a depth of 2566 m, just above the chalk reservoir. The maximum subsidence above the reservoir is approximately 6 m and is consistent with key references (Sulak and Danielsen 1989). These subsidence results indicate that the assumed values of the model parameters are reasonable with reference to existing subsidence observations.

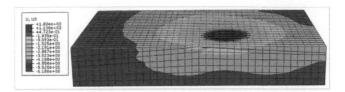

Figure 7.27. Subsidence field after pressure depletion near the wellbore (3D sectional view at a depth of 2566 m).

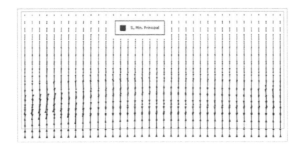

Figure 7.28. Initial orientation of minimum principal stress, YOZ plane view.

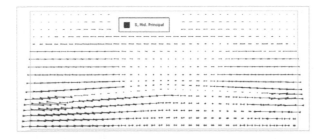

Figure 7.29. Initial orientation of medium principal stress, YOZ plane view.

This section provides the initial orientations of principal stresses. Because of the non-uniform distribution of pore pressure and non-uniform structure in the model, the initial orientations of principal stresses are non-uniform. Although the sectional view of the initial orientations of principal stresses in the Cartesian YZ plane shown in Fig. 7.28 through Fig. 7.31 appear to be regular, their sectional view of distribution in the Cartesian XY plane shown in Fig. 7.32 through Fig. 7.34 are significantly changed from their original status. The orientation of the medium and minimum principal stresses varies at the boundary that connects the reservoir with surrounding formations.

The numerical results of the directions of principal stresses after pore pressure depletion from 34 MPa to 10 MPa at the region shown in Fig. 7.26 are shown in Fig. 7.34 through Fig. 7.39.

As shown in Fig. 7.34 and Fig. 7.39, the directions of the minimum principal stress and that of the maximum principal stress after pressure depletion have changed significantly. A trajectory that is perpendicular to the direction of the minimum principal stress provides the best wellbore stability. Consequently, this change significantly influences the wellbore trajectory design. The wellbore trajectory at a field where pore pressure depletion has occurred, such as the Ekofisk field, should consider the geostructural characteristics, such as faults,

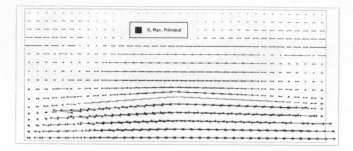

Figure 7.30.　Initial orientation of maximum principal stress, YOZ plane view.

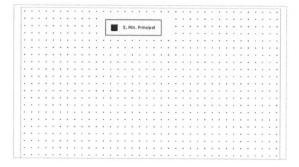

Figure 7.31.　Initial orientation of minimum principal stress, XOY plane view.

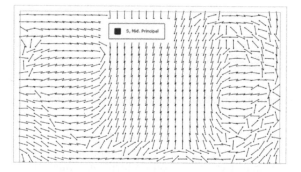

Figure 7.32.　Initial orientation of medium principal stress, XOY plane view.

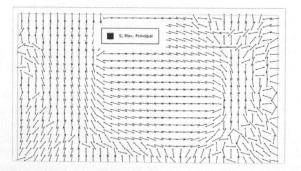

Figure 7.33.　Initial orientation of maximum principal stress, XOY plane view.

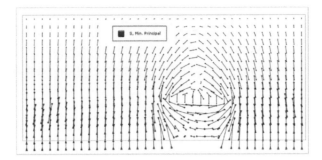

Figure 7.34. Orientation of minimum principal stress after pressure depletion, Cartesian YZ plane view.

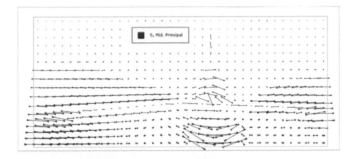

Figure 7.35. Orientation of medium principal stress after pressure depletion, YOZ plane view.

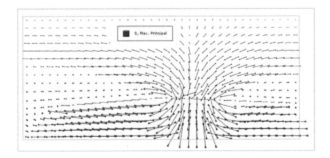

Figure 7.36. Orientation of maximum principal stress after pressure depletion, YOZ plane view.

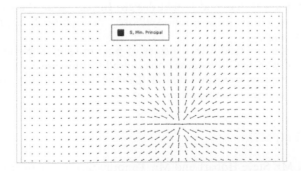

Figure 7.37. Orientation of minimum principal stress after pressure depletion, XOY plane view.

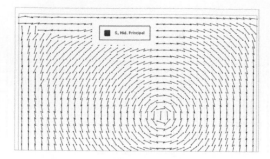

Figure 7.38.   Orientation of medium principal stress after pressure depletion, XOY plane view.

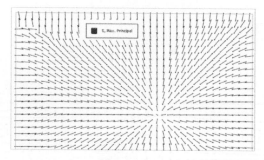

Figure 7.39.   Orientation of maximum principal stress after pressure depletion, XOY plane view.

and account for the variation of the principal stress directions caused by pore pressure depletion.

7.5   CONCLUSIONS

This chapter demonstrates that stress rotation can be caused by the presence of salt bodies and/or by pore pressure depletion. Theoretically, changes of hydrostatic stress at one material point will not change the principal directions of the stress tensor at this point. However, pore pressure depletion at a region can drastically change the principal stress direction of the field.

The numerical results presented in this work suggest the following:

- The variation of the orientations of principal stresses within a formation surrounding a salt body is important to the mud weight design. In particular, the stress ratio/tectonic factors at a salt-exit section of a subsalt wellbore will be difficult to predict empirically because of stress rotation.
- The trajectory design in a field with a history of pore pressure depletion should account for the variation of principal stress directions caused by pressure depletion. The direction of the minimum principal stress and that of the maximum principal stress after pressure depletion could be changed significantly.

ACKNOWLEDGEMENTS

The authors thank Mr. Steve Hobart and Mr. Timothy Sheehy for their constructive comments on the works related to this paper. They would also like to thank Dr. Joel Gevirtz for his discussion with the authors on this topic.

NOMENCLATURE

A	=	Model parameter in creep model
n	=	Model parameter in creep model
m	=	Model parameter in creep model
ε_z	=	Strain component in z-direction
σ_x	=	Effective stress component in x-direction
σ_y	=	Effective stress component in y-direction
σ_z	=	Effective stress component in z-direction
pp	=	Pore pressure gradient, ppg
ρ	=	Density, kg/m^3
TVD	=	True vertical depth, m
d	=	Cohesive strength used in Drucker-Prager criterion, Pa
β	=	Internal friction angle used in Drucker-Prager criterion, °
c	=	Cohesive strength, Pa
ϕ	=	Internal friction angle, °
$\bar{\sigma}^{cr}$	=	von Mises equivalent stress, Pa
$\dot{\bar{\varepsilon}}^{cr}$	=	Equivalent creep strain rate
E	=	Young's modulus, Pa
v	=	Poisson's ratio
k	=	Intrinsic permeability coefficient, Darcy, d
k_x	=	Intrinsic permeability coefficient in x-direction, Darcy, d
k_y	=	Intrinsic permeability coefficient in y-direction, Darcy, d
R	=	Initial void ratio
ρ_{clatic}	=	Density of clastic, kg/m^3
$\rho_{layer-2}$	=	Density of layer-2, kg/m^3
$\rho_{layer-3}$	=	Density of layer-3, kg/m^3
$\rho_{reservoir}$	=	Density of reservoir, kg/m^3
URL	=	Underground Research Laboratory
FEM	=	Finite element method

REFERENCES

Aburto, M. and Clyde, R.: The evolution of rotary steerable practices to drill faster, safer, and cheaper deepwater salt sections in the Gulf of Mexico. Paper SPE/IADC 118870 presented at the SPE/IADC Drilling Conference and Exhibition, Amsterdam, The Netherlands, 17–19 March, 2009.

Charlez, P.A., Bathellier, E., Tan, C. and Francois, O.: Understanding the present day in-situ state of stresses in the Cusiana field-Colombia. Paper SPE/ISRM 47208 presented at the SPE/ISRM Eurock, Trondheim, Norway, 8–10 July, 1998.

Cipolla, C.L., Hansen, K.K. and Ginty, W.R.: Fracture treatment design and execution in low-porosity chalk reservoirs. *SPEPO* 22:1 (2007), pp. 94–106.

Dassault Systems: *Abaqus Analysis User's Manual*, Vol. 3: Materials, Version 6.8, Vélizy-Villacoublay, France: Dassault Systems, 19.3.1-17–19.3.2-14, 2008.

Fredrich, J.T., Coblentz, D., Fossum, A.F. and Thorne, B.J.: Stress perturbation adjacent to salt bodies in the deepwater Gulf of México. Paper SPE 84554 presented at the Annual Technical Conference and Exhibition, Denver, CO, USA, 5–8 October, 2003.

Hickman, R.J.: *Formulation and implementation of a constitutive model for soft rock.* PhD Thesis, Virginia Polytechnic Institute and State University, Blacksburg, VA, USA, 2004.

Hunter, B., Tahmourpour, F. and Faul, R.: Cementing casing strings across salt zones: an overview of global best practices. Paper SPE 122079 presented at the Asia Pacific Oil and Gas Conference & Exhibition, Jakarta, Indonesia, 4–6 August, 2009.

Infante, E.F. and Chenevert, M.E: Stability of borehole drilled through salt formations displaying plastic behaviour. *SPE Drilling Engineering* 4:1 (1989), pp. 57–65.

Kristiansen, T.G.: Drilling wellbore stability in the compacting and subsiding Valhall field. Paper IADC/SPE 87221 presented at the IADC/SPE Drilling Conference, Dallas, TX, USA, 2–4 March, 2004.

Last, N.C. and McLean, M.R.: Assessing the impact of trajectory on wells drilled in an overthrust region. *JPT* 48:7 (1996), pp. 624–626.

Maia, A., Poilate, C.E., Falcao, J.L. and Coelho, L.F.M.: Triaxial creep tests in salt applied in drilling through thick salt layers in Campos basin-Brazil. Paper SPE/IADC 92629 presented at the SPE/IADC Drilling Conference, Amsterdam, The Netherlands, 23–25 February, 2005.

Martin, C.D. and Chandler, N.A.: Stress heterogeneity and geological structures AECL research. Whiteshell Laboratories, Pinawa Manitoba, Canada ROE1 L0 (1993), pp. 617–620.

Marinez, R., Shabrawy, M.E., Sanad, O. and Waheed, A.: Successful primary cementing of high-pressure saltwater kick zones. Paper SPE 112382 presented at the SPE North Africa Technical Conference & Exhibition, Marrakech, Morocco, 12–14 March, 2008.

Miskimins, J.L., Hurley, N.F. and Graves, R.M.: 3-D stress characterization from hydraulic fracture and borehole breakout data in a faulted anticline, Wyoming. Paper SPE 71341 presented at the Annual Technical Conference and Exhibition, New Orleans, LA, USA, 30 September–3 October, 2001.

Shen, X.P.: DEA-161 *Joint Industry Project to Develop an Improved Methodology for Wellbore Stability Prediction: Deepwater Gulf of Mexico Viosca Knoll 989 Field Area*. Halliburton Consulting, Houston, TX, USA, 18 August, 2009.

Sulak, R.M. and Danielsen, J.: Reservoir aspects of Ekofisk subsidence. *JPT* 41:7 (1989), pp. 709–716.

Teufel, L.W., Hart, C.M. and Sattler, A.R.: Determination of hydraulic fracture azimuth by geophysical, geological, and oriented-core methods at the multi-well experiment site, Rifle, CO. Paper SPE 13226 presented at the Annual Technical Conference and Exhibition, Houston, TX, USA, 16–19 September, 1984.

CHAPTER 8

Numerical analysis of casing failure under non-uniform loading in subsalt wells

Xinpu Shen

8.1 INTRODUCTION

Casing collapses in oil wells significantly affect the productivity and economical development of oilfields. These failures are usually associated with non-uniform loading, or with stress concentration as a result of following the influences:

- Adjacent or intruding salt bodies.
- Depletion, compaction, and subsidence of high-porosity reservoirs.
- Reactivation of faults or discontinuities.
- Complex tectonic active regions.

Examples reported in the literature include the deepwater subsalt in the Gulf of Mexico, offshore Angola, Brazil, north and west Africa, mature fields in the North Sea, and tectonic foothills in Colombia.

Willson and Fredrich (2005) reported the complexity of salt diapir styles in the Gulf of Mexico, which manifest a range of active, passive, and reactivate salt emplacement mechanisms. They observed that the domes of coastal Louisiana and of the Texas-Louisiana inner/mid shelf tend to be more complex. Diapir complexity further increases on the outer shelf to mid-slope trends where significant drilling problems have been observed in the sediments surrounding the diapir surface.

The most important characteristic of salt is its ability to creep under differential stress, which is a time-dependent phenomenon in two different time scales (transient and steady state response). The time-dependent behavior of the salt imposes a significant load on well casings; this load must be estimated in time and magnitude by using a numerical simulator to design the well construction accordingly.

The salt deformation rate (creep) is temperature dependent, which can dramatically increase the salt creep and the risk of drilling in a subsalt environment. These risks include the following:

- Tectonic instability in a active lateral salt deformation, thus inducing stress rotation and changing the stress field
- The presence of a rubble zone, depending on the salt geometry and the imposed effective stresses for salt emplacement or fluid migration
- Subsalt pressure regression, inducing either differential sticking or lost circulation
- Salt "gouge" at low-effective stress caused by a high lateral shear stress between the base of the salt and the adjacent formation.

Numerical experiments reported by Fredrich *et al.* (2003) have shown that the shear stress is highly amplified in certain zones in specific geometry adjacent to the salt-inducing anisotropy of horizontal stresses. In addition, the principal stresses are no longer vertical, and the horizontal and vertical stress may not be the maximum stress. For some geometries, vertical stress within and adjacent to the salt is not equal to the gravitational load (stress-arching effect occurs) before drilling or production; it behaves similarly to the overburden changes during depletion of compactable reservoirs.

Khalaf (1985) discussed alternative solutions for casing deformation occurring in some fields located near the salt body in northwest Africa. He adopts a mathematical model based on Lame's elastic solution of the thick-wall cylinder. His simulation, along with a triaxial lab test, indicates that salt has significantly lower mechanical properties than the adjacent sedimentary rocks. The in-situ strength of the salt reaches a maximum at a shallow depth (1524 m) and decreases in depth as the temperature increases; at 4572 m, it has a lower strength than at the surface. The non-uniform loading is amplified by the rate of creep, trajectory of the well (curvature), and the characteristics of the casing (mechanical and elastic properties and the geometry). The analyzed solutions include the use of thick-wall pipe and multiple casings packed with cement and high collapse resistant pipe.

El-Sayed and Khalaf (1992) discussed the effect of non-uniform loading in casing collapse when oil wells penetrated visco-elastic formations (salt). They presented a mathematical simulation for the single casing string and dual casing string under uniform/non-uniform-combined loading, assuming that collapse failure is caused by plastic yielding of the material in accordance with the von Mises criterion. Their results indicate that the reduction in loading resistance from uniform to non-uniform loading ranges from 70 to 80%, depending on the diameter/thickness ratio in the non-uniform loading condition. Consequently, increasing the pipe thickness is more effective than increasing the pipe grade to increase the collapse resistance of the casing. Non-uniform loading on casing can occur frequently without regional movement of the salt when an irregular borehole shape occurs because of the drilling fluid washing out the salt and inadequate filling of the annulus with cement. The salt then begins to flow into the wellbore in all directions at the same speed and causes uneven loading on the casing, which bends as a result of the salt creep. Considering that the movement of a layer of salt is caused by changes of the structure of the individual salt crystals when stressed, the deformation is the result of a intra-crystalline slip or dislocation. Consequently, the rate and severity of creep depends on the depth of burial, the stress differential between the wellbore and formation, temperature, mineralogical composition, water content, and the presence of impurities. Muecke and Mij (1993) propose an alternative solution to avoid casing failures related to squeezing-salt problems in the southern North Sea area. This approach consists of drilling with an oversaturated water-based mud to avoid washout and to enhance the quality of cementation.

The depletion of mature fields can also induce stress rotation and impose non-uniform high pressure on the casing through direct compaction of the reservoir rock, overburden fault, and bedding plane movement triggered by the reservoir compaction and in-well mechanical hot spots. Li *et al.* (2003) addressed the well failures at the Matagorda Island 623 field in the Gulf of Mexico after 16 years of production. Gravel pack screens lost their integrity at a compaction strain of 2.3%, and the compaction strain in cased and perforated completion began significant sand production at approximately 2.0 to 3.0%. However, casing damage in the overburden was reported to be less severe early on, but increased with tight spots and ovalized casing as the reservoir was depleted. The more frequent failures were observed in wells with poor cement jobs as a result of lost return during cementing which, like the washout in salt formations, leads to non-uniform loading. In addition, casing failures in the overburden were reported near the intersected fault, possibly resulting from the reactivation of this fault triggered by compaction and subsidence. Based on these experiences, the replacement of wells was designed to counteract the fault slip and to strengthen the pressure-seal capability of the wells. The annular space across the overpressured overburden was maximized to allow for fault movement.

Kristiansen (2004) discussed the compaction of chalk formation in Valhall field, which caused a seafloor subsidence near 5 m and a stress reorientation of approximately 90°. The subsidence continued at 0.25 cm/year. A large effort has been made to understand the great effect on wellbore stability that impose the surface changes followed by depletion, compaction, and subsidence, which also reactivate faults that impose high-shear loading in casing.

Da Silva *et al.* (1990) presented the results of a 3D finite element method (FEM) calculation using an elasto-plastic approach to explain the casing failure resulting from the subsidence of the Ekofisk reservoir in the North Sea. However, this model did not take into account the effect of discontinuities that severely affect the failures observed in the overburden.

Furui *et al.* (2009) reported a comprehensive modeling for both borehole stability and production liner deformation for inclined/horizontal wells in highly compacted chalk formations, including the effect of critical cavity dimensions caused by acid stimulation. They claim that the casing deformations in the low-angle section are related to the reservoir compaction, whereas the deformations in horizontal sections are induced by increased axial loading as a result of cavity deformation. Conversely, they found that the abnormal ductility of chalks after pore collapse around the borehole enhances borehole stability and reduces the risk of the liner deformation for openhole completions.

As previously mentioned, casing ovalization has also been detected in tectonically active fields, such as the foothills of Colombia, South America. Last *et al.* (2006) reported a significant rotation of stresses in the tectonically active foothills of Cusiana field in Colombia, in which the combined effect of the loading and structure (faults and dips) produce shear stresses in the overburden. These shear stresses resulted in a rotation of principal stresses away from the vertical and horizontal directions, leading to significant wellbore stability problems while drilling. However, this condition also imposed non-uniform loading on the casing string during production that resulted in casing ovalization. Patillo *et al.* (2003) reported the results of numerical and experimental studies to account for the effect of non-uniform loading on the conventional collapse resistance of casing. They also addressed the physics of the apparent inconsistency between the field observations and the dramatic reduction in collapse resistance when ovalization increases. In addition, they performed a numerical simulation, validated by laboratory testing, proving that ovalization, resulting from non-uniform formation loading, can lead to unacceptable cross-sectional deformation, decrease the resistance of a cross section to more conventional loading by fluid pressure differential and a lower manufacturing ovality, and increase conventional collapse resistance. It can also decrease the imposed ovalization value at which non-uniform loading begins to reduce collapse resistance.

Based on this improved understanding, Last and McLean (1996) created an operational strategy for managing the effect of casing ovalization in this region. They developed new equations that relate collapse resistance to ovality for supported (good cement) and unsupported (poor cement) casing. Their results indicate that the progression of ovalization is characterized by an early time response in which the ovalization rapidly reaches a threshold, followed by a much lower rate that appears to be constant, at least for the period over which data are available.

Because of their significance to the industry, salt and casing failure in Paradox basin have been studied since the 1950s (MacAlister 1959; Tsai and Silva 1987; Russell *et al.*, 1984) The salt layers are very thick, which increases the difficulty of drilling through them.

This chapter uses a casing example of a subsalt wellbore as the object of finite element modeling and analysis. Plastic deformation caused by non-uniform properties of the cement ring was simulated with various values of Young's modulus at different parts of the cement ring. As an enhancement measure, the structure of double casing separated by cement rings, rather than single casing, was adopted. The mechanical behavior of this enhancement measure was numerically simulated.

This example is provided for illustrative purposes; all data used in this example are fictionalized or modified on the basis of some practical cases in engineering, but none of data is directly from any commercial project.

8.2 FINITE ELEMENT MODEL AND ANALYSIS OF CASING INTEGRITY

A key issue in building a model for numerical casing integrity estimates is a proper understanding of the mechanism that controls the generation of non-uniform load from salt to casing. As introduced in the last section, various mechanisms underlie the casing failure phenomena. Among those factors, the imperfection of completion is believed to be the most significant factor. In general, the manufacturing fault existing within casing itself is neglected. Completion imperfection will result in a non-uniform distribution of mechanical properties, such as stiffness, within the concrete rings. If the stiffness of the concrete ring is not uniform, when salt compression occurs to the concrete as a result of the inward creep deformation, the resistant force generated from the concrete ring is consequently non-uniform. The compressional force generated will be greater at the point with greater stiffness values than that generated at the point with lesser stiffness values. In this case, even with uniform creep deformation of the surrounding salt, the force applied on the outer surface of casing by the salt is not uniform. Therefore, completion imperfection will be numerically simulated in the following sections, and the casing integrity will then be investigated with a model that has imperfect completion. The concrete ring used in this model has two parts, which have different stiffness properties.

In addition, the extent of completion imperfection must be determined, such as the range of the concrete ring in which the stiffness abnormality occurs. This should be determined with a type of non-destructive detection device, such as sonic detector. In this example, both the value of stiffness and the range of concrete where imperfection occurs are assumed. The values assumed here are physically possible; consequently, it is reasonable to use them in this illustrative calculation.

Because of the large differences between the field scale and the casing section scale in modeling, it has been difficult, if not impossible, to combine these models in the past. In fact, existing examples of numerical analyses of casing failure were either performed at a reservoir scale without direct coupling to the field scale behaviors, or were performed at a much larger scale, which sacrificed much of the necessary modeling resolution.

Submodeling techniques are used to accommodate the field-to-casing-section scale discrepancy. Using this approach, a highly inclusive field scale analysis can be linked to a very detailed casing stress analysis at a much smaller scale.

The position of the submodel has been chosen at the lower salt exit; this is the position where casing failure is likely to occur because of the complicated stiffness distribution around the borehole. The stress distribution is also complicated because of the change of pore pressure across the interface between the salt and the reservoir.

The following sections present the numerical analysis at a field scale, followed by an analysis of the submodel at a casing-section scale.

8.2.1   *Numerical analysis of global model at field scale*

8.2.1.1   *Model geometry*
The geometry of the model used in this chapter is similar to the model used in previous chapters, and is briefly restated in this section.

Fig. 8.1 provides the salt geometry, with a maximum outer radius equal to 7 km; Fig. 8.2 shows the field scale model. The total model thickness is 10 km on the left side and 9.6 km on the right side, which reflects a variation of surface elevation from left to right. Its width and length are 10 km, respectively.

8.2.1.2   *Material models*
A simplified model and four kinds of materials have been adopted, including the upper formation, lower formation, formation surrounding the salt, and the salt body. Table 8.1 includes a list of the parameters.

This analysis uses the modified Drucker-Prager yielding criterion. The values of material properties are listed in the following paragraphs. The values of strength parameters for salt adopted by the modified Drucker-Prager model are $d = 4$ MPa, $\beta = 44$ which correspond to values in the Mohr-Coulomb model as $c = 1.25$ MPa, $\phi = 25°$.

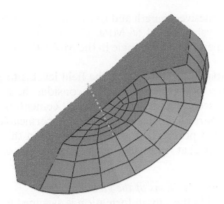

Figure 8.1. Salt geometry: half salt body.

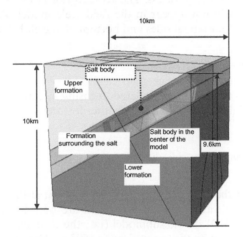

Figure 8.2. Model geometry: profile of the field model.

Table 8.1. Values of material parameters.

	ρ/kg/m³	E /Pa	v
Upper and surrounding formation	2300.	1.9×10^{10}	0.25
Lower formation	2400.	1.9×10^{10}	0.25
Salt	2150.	1.8×10^{10}	0.25

The following creep law in Eq. 8.1 (Dassault Systems, 2008) is adopted:

$$\dot{\bar{\varepsilon}}^{cr} = A\left(\bar{\sigma}^{cr}\right)^{n} t^{m} \tag{8.1}$$

where $\dot{\bar{\varepsilon}}^{cr}$ represents the equivalent creep strain rate, $\bar{\sigma}^{cr}$ represents the von Mises equivalent stress, and t is total time variable. *A*, *n*, and *m* are three model parameters, which are given by the following values:

$$A = 10^{-21.8}, \quad n = 2.667, \quad m = -0.2 \tag{8.2}$$

For the formation, the cohesive strength and frictional angle of the Drucker-Prager model are given by the following values: $d = 1.56$ MPa, $\beta = 44°$. These two values correspond to values of cohesive strength and frictional angle in the Mohr-Coulomb model as $c = 0.5$ MPa, $\phi = 25°$.

The purpose of the global model analysis at the field level is to provide a set of accurate boundary conditions for the local casing section. To consider the salt exit section of the wellbore, which is the weakest section and where imperfect cementing can occur, it is assumed in this study that only the part of the lower formation is permeable. Consequently, coupled analyses for deformation and porous flow were made only in this lower formation region. Other parts of the global model at field scale are assumed to be nonpermeable.

8.2.1.3   *Loads and boundary conditions of the global model*

The initial pore pressure within the subsalt formation is assumed to be 47 MPa. As shown in Fig. 8.3, loads applied to the model at the field scale include the self-gravity of formations, which is balanced with the initial geostress. The geometry of borehole and mud weight pressure within the borehole will not appear in the field scale model. Zero-displacement constraints are applied to the four lateral sides and bottom of the global model.

8.2.1.4   *Numerical results of global model*

The global model analysis connects the geometrical factors of the salt body to the submodels. Because numerical results are not directly related to casing integrity, the details of numerical results of the global model are omitted in this chapter for brevity.

8.2.2   *Submodel and casing integrity estimate*

8.2.2.1   *Model geometry*

A submodel was built for the wellbore section of the salt exit at the lower center of the salt body (see Fig. 8.4). The salt and reservoir formation are both included in this submodel. The true vertical depth (TVD) of the submodel (i.e., the distance from center of the submodel to its right vertical point on the top surface) is 3595 m. The upper part of the submodel is salt, and the lower part is the reservoir formation, which has a pore pressure of 47 MPa.

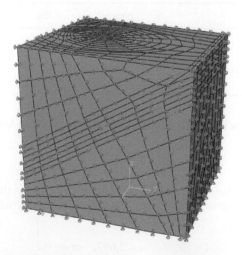

Figure 8.3.   Loads and boundary conditions of the model at field scale.

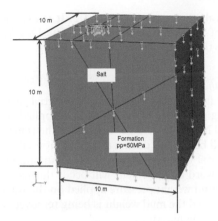

Figure 8.4.   Model geometry: profile of the submodel.

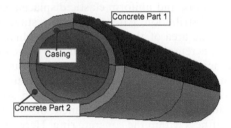

Figure 8.5.   Illustration of casing and concrete ring.

Table 8.2.   Values of material parameters for the two parts of the concrete ring.

Concrete ring	E/Pa	ν	d/Pa	$\beta/°$
Part 1	3×10^{10}	0.2	15×10^{6}	40
Part 2	1×10^{10}	0.2	5×10^{6}	25

The wellbore axis has an inclination angle of 30° in the vertical direction, which is perpendicular to the salt-reservoir interface.

The casing and concrete ring are included in the submodel. As shown in Fig. 8.4, the inner and outer casing radii are 0.0752 m and 0.0889 m, respectively. The outer radius of the concrete ring is 0.108 m. A domain of formation surrounding the wellbore with a radius of 0.8 m is assumed to be the region that will be influenced by skin effect.

8.2.2.2   *Material models*
The casing material is assumed to be elasto-plastic and to have the following elastic and initial plastic strength parameters values:

$$E = 2 \times 10^{11}\,\text{Pa}, \quad \nu = 0.3, \quad \sigma_s = 3 \times 10^{8}\,\text{Pa} \tag{8.3}$$

Other details of casing, such as the type of steel, have been neglected here.

To simulate cementing imperfection, the concrete ring was been separated into two parts, as shown in Fig. 8.5. As listed in Table 8.2, different values for the mechanical properties are assumed for these two parts of the concrete ring. Part 1 is assumed to have greater stiffness and strength parameters values than those of Part 2. It is assumed that the sectional area of Part 1 is one quarter of the entire sectional area. The Drucker-Prager model has been adopted for characterizing the materials of the concrete ring.

8.2.2.3 *Loads specific to the submodel*

The loads of the submodel include self-gravities and the initial geostress field, which balances the gravity. For the drilling stage, a safe mud weight is applied on the inner surface of the borehole, which is set at 70 MPa, and will be removed in the cementing stage.

Mud weight pressure will be applied during the simulation of the drilling process. Drilling as a disturbance to the initial geostress field will cause an additional geostress load to the wellbore. Consequently, inward displacement will occur during the simulation of the drilling process, even with the mud weight pressure applied on the wellbore surface. Particularly, when the casing is being set and the mud weight is being removed, additional geostress loads will apply to the casing and concrete ring.

Load caused by salt creep will be applied to the concrete ring and casing. This creep load develops with time. In an ideal state, the mechanical properties of the concrete and casing are normally in an elastic status, and uniform elastic displacement could occur, followed by a stable uniform stress redistribution. After the primary stage of deformation development, displacement within the area of the wellbore will gradually cease. However, when imperfection exists in the concrete ring, creep deformation will cause a non-uniform stress distribution within the concrete and will cause additional non-uniform stress distribution to the casing. In some instances, this non-uniform stress will result in inward plastic casing failure. The best means of determining the loading condition associated with salt creep-related failure is to have the original in-situ measurement of the shape of the casing along with the distribution of stiffness within the concrete ring. Here, the calculation adopts the variation of pore pressure in the area of the wellbore to generate the displacement of salt. Consequently, non-uniform deformation occurs within the concrete ring. The numerical results indicate that the effect of this kind of load is equivalent to the expected salt creep displacement.

Additional loads include the pore pressure within the formation and the pressure on the inner surface of the casing, which is equal to the pressure in the formation. Skin effect will be accounted for through a reduction of pore pressure within the formation around the wellbore. Various values of pore pressure have been assumed to simulate this skin effect-related variation in the study.

The boundary conditions for the submodel are derived directly from the numerical results obtained with the global model.

8.2.2.4 *Numerical results of the submodel: Stress distribution around the borehole before cementing*

In this study, a stress analysis with a safe mud weight was simulated first. At this stage, the concrete ring and casing do not appear. Fig. 8.6 through Fig. 8.8 show the distribution of principal stresses around the borehole. The green circle in the center represents the position of casing inner surface (sectional view). The depth of the section view is chosen in the salt body near the salt exit.

Because the wellbore investigated is inclined, the sectional view is perpendicular to the axis of the wellbore. Thus, it is an inclined sectional view. The vertical gravity component has an intersection angle with the sectional view. Fig. 8.6 shows that the vertical stress distribution is basically uniform. In addition, the medium principal stress distribution is nearly uniformly around the borehole (Fig. 8.7). In Fig. 8.8, distributions of the vectors are not uniform; intervals between vectors vary with central angles, which indicate that the maximum stress distribution around the borehole is not uniform to some extent.

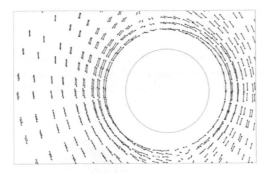

Figure 8.6.   Distribution of minimum principal stress around the borehole after drilling, but before cementing.

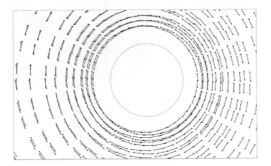

Figure 8.7.   Distribution of medium principal stress around the borehole after drilling, but before cementing.

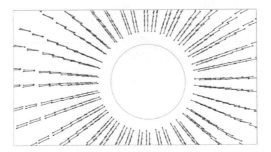

Figure 8.8.   Distribution of maximum principal stress around the borehole after drilling, but before cementing.

Casing integrity has been investigated with assumptions of cementing imperfection and pore pressure changes related to skin effect.

8.2.2.5   *Numerical results of submodel: Stress distribution within the concrete ring and casing*

The pore pressure changes were found to result from skin effect, which is too weak to cause casing failure. However, skin effect can cause a non-uniform distribution of pore pressure within the formation during the pore pressure depletion process/oil production process.

A variation of permeability coefficients will result in a non-uniform porous flow rate within the formation; however, non-uniform pressure will disappear with time if no flow develops at the model boundary.

When the depletion of pore pressure within the formation was set from 47 MPa to 35 MPa in this calculation, casing failure occurred. Fig. 8.9 shows the distribution of the equivalent plastic strain, which is a scalar. Plastic deformation occurs at a small portion of casing on its upper end, and the z-axis is upward. Because the model (including geometry, material,

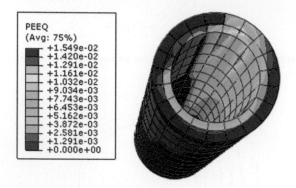

Figure 8.9.   Distribution of plastic strain within casing.

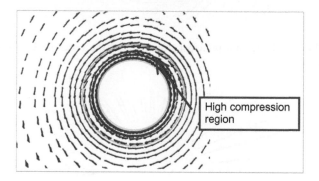

Figure 8.10.   Distribution of minimum principal stress around borehole after cementing.

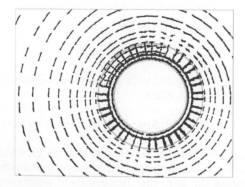

Figure 8.11.   Distribution of medium principal stress around borehole after cementing.

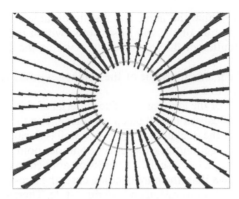

Figure 8.12.   Distribution of the maximum principal stress around the borehole after cementing.

and loads) is not symmetrical, the plastic deformation of the casing is not symmetrical. The maximum plastic deformation occurs where it connects to the concrete ring, which has a higher Young's modulus value.

This occurrence indicates that casing failure will occur if imperfect cementing occurs across a large percent of the sectional area of the concrete ring.

An alternative case has been tested in which the values of the parameters are alternatively asserted: the small portion of concrete ring section has weak strength, and the large portion has sound strength. The casing failure, therefore, does not occur.

Fig. 8.10 through Fig. 8.12 provide the distributions of principal stresses after cementing. The concrete ring and casing have been set to their positions. As shown in Fig. 8.10, the high compression region was formed at the position where the concrete had a high Young's modulus value and a higher strength than that of the left part. The squeezing load from the salt formation was transferred to the casing in this region before other parts, which consequently caused the plastic failure of the casing.

Fig. 8.11 shows the non-uniform distribution of the medium principal stress. In the upper-left area near Part 1 of the concrete ring, orientations of the medium principal stress at some points are distributed in a hoop direction within the concrete ring along with radiant distribution of the medium principal stress at other points.

8.3   NUMERICAL RESULTS OF ENHANCEMENT MEASURE

One of the enhancement measures for well sections at the salt base was the use of a dual casing structure. This measure drastically reduces the risk of non-uniform distribution of mechanical properties within concrete rings. Consequently, the stress concentration within the inner casing is significantly reduced.

Fig. 8.13 shows a model of a well section completed with a casing-cement-casing2-cement alternative. In this design, the thickness of the inner casing is thinner than that of the previous casing because there are two casings in this structure. The thickness of the casing shown in Fig. 8.13 is approximately half the thickness of the casing shown in Fig. 8.9.

It is assumed that concrete Part 1 and Part 2 have different mechanical properties values, as listed in Table 2. To simulate a possible situation in which the non-uniform distribution of concrete properties was caused by completion imperfection, positions of concrete Part 1 in two concrete rings are positioned as those shown in Fig. 8.13. Other parts of the model, such as boundary conditions and loads, use the same settings as those used in the previous model with single casing.

Fig. 8.14 and Fig. 8.15 show the numerical results obtained with the given structure. Fig. 8.14 shows the distribution of the von Mises stress within the model. Fig. 8.15 shows the selected view of distribution of the von Mises stress within the dual casings and concrete rings only.

Fig. 8.15 shows that the maximum value of the von Mises stress is 202 MPa, which is far less than the strength given in Eq. 8.3.

Fig. 8.16 shows the mesh of the inner casing. The material of this casing is under a bending stress status under the load caused by salt creep squeezing. For the pure bending load case, the stress distribution along the thickness of the casing wall is linear: it varies from tension to compression in the radial direction/thickness direction under point loading. For the case of loading with salt creep squeeze, because the compression applied to the casing is high and the non-uniform extent is far less than the pure bending case, the stress status within the casing is actually in a state of mixed compression and bending state. Therefore, the stress within the casing wall varies in the thickness direction from a small compressive stress value to a high value in a linear pattern.

Because of the linear distribution of the stress in the thickness direction, the mesh for the casing requires multiple discretization along the thickness of the casing wall. This will make the computational cost rather expensive. More significantly, it will challenge the computer capacity because of the large amount of nodal freedom required.

In Fig. 8.16, the mesh of the inner casing has been discretized by two layers of elements in its thickness direction. This is the minimum number required for a reasonable stress solution.

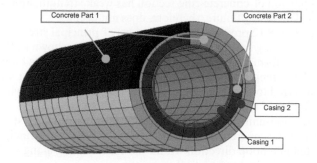

Figure 8.13.　Casings and concrete rings.

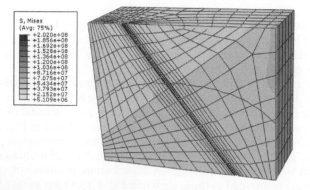

Figure 8.14.　Distribution of the von Mises stress within the model.

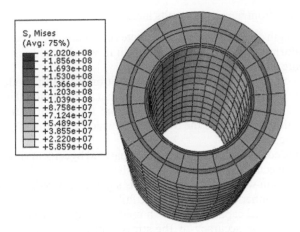

Figure 8.15. Selected view of distribution of the von Mises stress within the dual casings and concrete rings.

Figure 8.16. Mesh of the inner casing.

Even so, the stress solution obtained with this mesh will be significantly less than its actual value. Consequently, we have adopted a rather small value of strength given in Eq. 8.3 for this calculation.

If the computer capacity permits, the use of additional layers in the thickness direction will result in achieving a better stress solution. The use of quadratic elements will also significantly improve the accuracy of the numerical stress solution

In practice, the estimates of the stress distribution within casing are more accurate when analytical means are used for the estimate; special attention should be paid to the variation of the stress along thickness/radial direction.

8.4 CONCLUSIONS

Stress analyses within a salt base well section were performed with a 3D FEM. Casing failure caused by cementing imperfection was investigated numerically. A multi-scale numerical technique was adopted in the analysis.

Conclusions derived from numerical results include the following:

- Non-uniform strength distribution of the concrete ring caused by imperfect cementing will result in non-uniform loading to the casing within it, which could result in casing failure when pore pressure depletion and salt creep occur.
- A high-compression region was formed at the position where the concrete had a higher Young's modulus value and a greater strength than that of the material in places other than the high-compression region. The squeeze load from the salt formation was transferred to the casing in this region before the other parts, which consequently caused the plastic failure of the casing.
- By using a structure consisting of double casing separated by cement rings, the non-uniform distribution of the mechanical properties within the concrete rings is eased. Consequently, the stress concentration within the inner casing is significantly reduced. This measure can effectively protect the inner casing from plastic failure.
- Limited by computational capacity, the mesh used in the discretization of casing is rather coarse. Consequently, the accuracy of the stress solution within the casing is not perfect. If computer capacity permits, the use of additional layers in the thickness direction will achieve a better stress solution. The use of quadratic elements will also significantly improve the accuracy of numerical stress solutions.

Cementing is a complicated multiphase process. In a complete cementing process, major processes consist of cement filling and cement solidification in which a chemical process is also involved. In this process, the function of the mud weight pressure was eventually replaced by a concrete ring. A certain amount of formation deformation could occur before the final solid concrete ring was formed.

The results described in this chapter were obtained at a primary stage in which the simulation of this complicated process is performed with a simplified model. Additional work will be performed to investigate the cementing process, which can be influenced by the contents of cement materials and is the resource of non-uniform distribution of material properties of the concrete ring.

ACKNOWLEDGEMENTS

Thanks are due to Dr. Arturo Diaz, Mr. Steve Hobart, and Mr. Timothy Sheehy for their constructive comments on the works related to this paper, and to Dr. Joel Gevirtz for his helpful discussion with the author on this topic.

Partial financial support from China National Natural Science Foundation through contract 10872134, support from Liaoning Provincial Government through contract RC2008-125 and 2008RC38, support from the Ministry of State Education of China through contract 208027, and support from Shenyang Municipal Government's Project are gratefully acknowledged.

NOMENCLATURE

A = Model parameter in creep model
n = Model parameter in creep model
m = Model parameter in creep model
d = Cohesive strength used in Drucker-Prager criterion, Pa
β = Internal friction angle used in Drucker-Prager criterion, °
c = Cohesive strength, Pa
ϕ = Internal friction angle, °
E = Young's modulus, Pa
v = Poisson's ratio
ρ = Density, kg/m^3
σ_s = Initial strength used in von Mises criterion, Pa

K = Ratio parameter used in Drucker-Prager criterion
$\bar{\sigma}^{cr}$ = Von Mises equivalent stress, Pa
$\dot{\bar{\varepsilon}}^{cr}$ = Equivalent creep strain rate
p_{mw} = Safe mud weight pressure, Pa
FEM = Finite element method
TVD = True vertical depth

REFERENCES

Dassault Systems: Abaqus Analysis User's Manual, Vol. 3: *Materials, Version 6.8*, Vélizy-Villacoublay, France: 19.3.1-17–19.3.2-14, 2008.

Da Silva, F.V., Bebande, G.F., Pereira, C.A. and Plischke, B.: Casing collapse analysis associated with reservoir compaction and overburden subsidence. Paper SPE 20953 presented at Europec 90 held in The Hague, The Netherlands, 22–24 October, 1990.

El-Sayed, A.A.H. and Khalaf, F.: Resistance of cemented concentric casing strings under nonuniform loading. Paper SPE 17927 in *SPE Drilling Engineering* 7:1 (1992), pp. 59–64.

Fredrich, J.T., Coblentz, D., Fossum, A.F. and Thorne, B.J.: Stress perturbation adjacent to salt bodies in the deepwater Gulf of México. Paper SPE 84554 presented at the Annual Technical Conference and Exhibition, Denver, CO, USA, 5–8 October, 2003.

Furui, K., Fuh, G-F., Abdelmalek, N. and Morita, N.: A comprehensive modeling analysis of borehole stability and production liner deformation for inclined/horizontal wells completed in highly compaction chalk formation. Paper SPE 123651 presented at the Annual Technical Conference and Exhibition, New Orleans, LA, USA, 4–7 October, 2009.

Khalaf, F.: Increasing casing collapse resistance against salt-induced loads. Paper SPE 13712 presented at the Middle East Oil Technical Conference and Exhibition, Bahrain, 11–14 March, 1985.

Kristiansen, T.G.: Drilling wellbore stability in the compacting and subsiding Valhall Field. Paper IADC/SPE 87221 presented at the IADC/SPE Drilling Conference, Dallas, TX, USA, 2–4 March, 2004.

Last, N.C. and McLean, M.R.: Assessing the impact of trajectory on wells drilled in a overthrust region. Paper SPE 30465 in *JPT* 48:7 (1996), pp. 624–626.

Last N.C., Mujica, S., Pattillo, P.D. and Kelso, G.: Evaluation, impact, and management of casing deformation caused by tectonic forces in the and near foothills, Colombia. Paper SPE 74560-PA in *SPE Drilling and Completion* 21:2 (2006), pp. 116–124.

Li, X., Mitchum, F.L., Bruno, M., Patillo, P.D. and Willson, S.M.: Compaction, subsidence, and associated casing damage and well failure assessment for the Gulf of Mexico Shelf Matagorda Island 623 Field. Paper SPE 84553 presented at the Annual Technical Conference and Exhibition, Denver, CO, USA, 5–8 October, 2003.

MacAlister Jr., R.S.: How oil wells are completed in the Paradox basin. Paper SPE 1220-G in *JPT* 11:8 (1959), pp. 18–22.

Muecke, N.B. and Mij, B.V.: Heated mud system: a solution to squeezing-salt problems. Paper SPE 25762 presented at SPE/IADC Drilling Conference, Amsterdam, The Netherlands, 23–25 February, 1993.

Pattillo, P.D., Last, N.C. and Asbill, W.T.: Effect of non-uniform loading on conventional casing collapse resistance. Paper SPE/IADC 79871 presented at the SPE/IADC Drilling Conference, Amsterdam, The Netherlands, 19–21 February, 2003.

Russell, J.E., Carter, N.L. and Handin, J.: Laboratory testing for repository characterization. Paper 84-1188 presented at the 25th U.S. Symposium on Rock Mechanics (USRMS), Evanston, IL, USA, June 25–27, 1984.

Shen, X.P.: DEA-161 *Joint Industry Project to Develop an Improved Methodology for Wellbore Stability Prediction: Deepwater Gulf of Mexico Viosca Knoll 989 Field Area*. Halliburton Consulting, Houston, Texas, USA, 18 August, 2009.

Tsai, F.C. and Silva, W.: Basement rock faulting as a primary mechanism for initiating major salt deformation features. Paper ARMA 87-0621 presented at the 28th U.S. Symposium on Rock Mechanics (USRMS), Tucson, AZ, USA, 29 June–1 July, 1987.

Willson, S.M. and Fredrich, J.T.: Geomechanics considerations for through- and near-salt well design. Paper SPE 95621 presented at the Annual Technical Conference and Exhibition, Dallas, Texas, USA, 9–12 October, 2005.

CHAPTER 9

Numerical predictions on critical pressure drawdown and sand production for wells in weak formations

Xinpu Shen

9.1 INTRODUCTION

Sand production is an important issue affecting oil production; it can cause serious problems in oil flow within a sand reservoir as well as to the proper functioning of the oil production equipment. Furthermore, for wells with openhole completion, sand production can cause cavity enlargement and further collapse of the borehole. The major cause of sand production is the material instability in poorly consolidated and unconsolidated formations. As a kind of material instability, it was proved that material plasticity plays an important role in the process of sand production (Papamichos and Stavropoulou 1998; Papamichos and Malmange 2001; van den Hoek *et al.*, 1996). In the last twenty years, the investigation and application of plasticity-based methods for the prediction of sand production have been reported by various researchers. Wang and Lu (2001) introduced their work on the relationship between the onset of sand production and equivalent plastic strain by using a coupled reservoir-geomechanics model. Yi *et al.* (2004) proposed their sanding onset-prediction model, which can analytically calculate the critical pressure drawdown by using a set of material-flow parameters. Oluyemi and Oyeneyin (2010) presented their analytical model for the prediction of critical pressure drawdown that is based on the Hoek-Brown failure criterion, rather than on the Mohr-Columb criterion.

The elastoplastic consolidation analysis of a perforation tunnel around a wellbore is the theoretical basis of a sand production prediction. In recent years, various researchers have contributed their efforts to this topic (Oluyemi and Oyemeyin 2010; He 2005; Liu 2005; Zhuang *et al.*, 2005; Dassault Systems 2008b).

In general, the calculation for sand production is performed in a plasticity-based manner. Specifically, if no plasticity occurs around the borehole and perforation tunnel, then the risk of sanding is zero. If plastic strain occurs, there is a risk of sanding. The amount of risk depends on the amount of equivalent plastic strain at each material point. If plastic strain occurs on a large area around the borehole (for openhole completion) and around the perforation tunnel, the sanding potential is high. Consequently, the calculation must be performed with the 3-dimensional finite element method (3D FEM).

The problems of flow within porous underground media were clarified by He (2005) with a detailed review of the coupling between multi-phase flow and rock deformation. Liu (2005) investigated sand production under an isotropic horizontal geostress state and presented an equation for the position of the inner surfaces of shots.

The critical value of pressure drawdown (CVPDD) is a key parameter that controls the production rate for most of the wells in weak formations and depends on the form of completion. According to references and engineering observations, the critical value of equivalent plastic strain depends on the following:

- Formation strength properties.
- Formation pore pressure.
- Geostress tensor, both mean stress and stress deviator.

- Grain size.
- Formation thickness and other geometrical parameters.

This chapter presents two cases in which the 3D FEM was applied to predict the CVPDD in two wells among which one with openhole completion and another well with casing completion, respectively. The CVPDD was chosen on the basis of maximum plastic strain obtained with a 3D numerical calculation for a given well. The critical value of the equivalent strain chosen was 1% for wells with openhole completion and 10% for wells with casing completion. A fully coupled, poro-elastoplastic model (see Zhuang *et al.,* 2005; Dassault Systems 2008a) was adopted to simulate the porous flow that occurred simultaneously with matrix inelastic deformation. A submodeling technique (Shen 2010) was adopted to address the discrepancy between the scale of the oil field and that of the wellbore section. A full-field scale, which is 3 km in depth and 5 km in width and length, was used to establish a geostress field. The pressure drawdown and poro-elastoplastic behavior of the formation near the wellbore was calculated with a submodel, which has a diameter of 7 m. Based on the results provided by the 3D finite element calculation, critical pressure drawdown before the onset of sanding was predicted; a CVPDD of 400 psi was proposed for a well with openhole completion, and a CVPDD of 600 psi was proposed for a well with casing completion.

Section 9.5 presents the calculation of the amount of sand volume produced during a four-day period. The erosion of the perforation tunnel was visualized.

9.2 MODEL DESCRIPTION AND NUMERICAL CALCULATION

Submodeling techniques are used to accommodate the field-to-casing-section scale discrepancy. The concept of the submodeling technique includes using a large-scale global model to produce the boundary conditions for a smaller scale submodel. In this way, the hierarchical levels of the submodel are not limited. This approach enables a highly inclusive field-scale analysis to be linked to a very detailed casing-stress analysis at a much smaller scale. The benefits are bidirectional, with both the larger and smaller scale simulations benefiting from the linkage.

9.2.1 *Numerical calculation with global model*

Fig. 9.1 shows the field-scale model.

The total depth of the model is 3000 m, the width is 5000 m, and the length is 5000 m. The model uses four vertical layers of overburden; the first layer is 1000 m, the second layer is 500 m, the depth of third layer varies from 435 m on the left and 800 m on the right (as shown

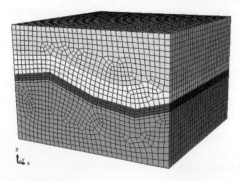

Figure 9.1.   Geometry of the field model.

in Fig. 9.2), and the thickness of bottom layer varies between 900 and 1265 m. The reservoir layer that ranges from 50- to 150-m thick is located in the lower part of the model, as shown in Fig. 9.2.

The two wells are both vertical wells, as shown in Fig. 9.2. The true vertical depth (TVD) of the target reservoir of well #1 is 1622 m, and the TVD for well #2 is 2560 m.

9.2.1.1   *Values of material parameters*
A simplified model and two kinds of materials were adopted, including the upper formation, lower formation, and the reservoir layer at well #1 and the reservoir section at well #2. For simplicity, the density, Young's modulus, and Poisson's ratio have been given the same value. The reservoir was assumed to be permeable with a permeability set at 500 mD where well #1 is located and 1000 mD where well #2 is located; other formations were assumed to be non-permeable. Table 9.1 lists the parameters.

For well #1, the formation properties of the reservoir section include the following: the void ratio is set at 0.33, porosity is 0.25, permeability is 500 mD, the frictional angle is set at 25°, and the cohesive strength is chosen to be 3.5 MPa.

The formation properties of the reservoir section for well #2 are as follows: the void ratio is set at 0.35, porosity is 0.26, permeability is 1000 mD, the frictional angle is set at 25°, and cohesive strength is chosen to be 0.5 MPa.

The purpose of the global model analysis at the field level is to provide a set of accurate boundary conditions for the local casing section. To simplify the calculations without sacrificing the accuracy of the problem description, it is assumed in this study that only the part of the reservoir formation is permeable. Consequently, a coupled analysis for deformation and porous flow has been made only in this region. Other parts of the global model at field scale are assumed to be nonpermeable.

9.2.1.2   *Loads and boundary conditions of the global model*
The initial pore pressure within the reservoir is assumed to be 20.5 MPa at the location of well #1 and 38 MPa at the location of well #2. As shown in Fig. 9.3, the loads applied to the model at the field scale include seawater pressure and the self-gravity of the formations and salt, which is balanced with the initial geostress. Mud weight pressure will not appear

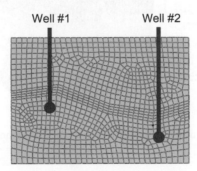

Figure 9.2.   Position of the wells within the field model.

Table 9.1.   Values of material parameters.

ρ/kg/m^3	E/Pa	ν
2200.0	1.0×10^{10}	0.34

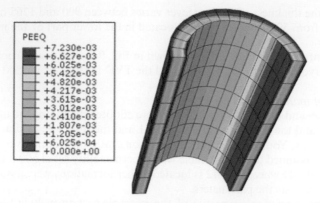

Figure 9.3.   Boundary conditions of the model at field scale.

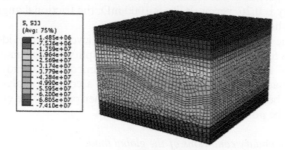

Figure 9.4.   Numerical results: distribution of von Mises stress.

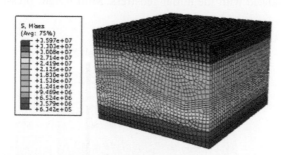

Figure 9.5.   Numerical results: distribution of the vertical stress component S33.

in the field-scale model. Zero-displacement constraints are applied to the four lateral sides and bottom.

9.2.1.3   *Stress pattern*

The stress regime of the field is a regular stress pattern: the vertical stress component is the maximum stress component and no obvious tectonic stress effect exists. It is assumed that the tectonic factor, tf, which indicates the extent of tectonic effect, is 0.25, and the effective stress ratio, k0, which is the ratio between the effective vertical stress component and the effective minimum horizontal stress, is 0.67. The Y direction, which corresponds to the north, was chosen as the orientation of the maximum horizontal stress component.

9.2.1.4 *Numerical results of global model*
The purpose of the global model analysis is to determine the positions at which maximum strain occurs and, thus, to estimate the casing integrity at those positions by using a submodeling technique. In addition, the numerical results of the global calculation at a field scale will provide boundary conditions for the submodels adopted in the analysis.

Fig. 9.7 shows the distribution of the vertical displacement component, which represents subsidence when it is negative. All three displacement components have been calculated and will be used as boundary conditions for the submodels used in the following sections.

9.3   CASE 1: PREDICTION OF CVPDD FOR A WELL WITH OPENHOLE COMPLETION

The task is to calculate the CVPDD for well #1 with an openhole completion. Poroelastoplastic calculations have been performed with submodels.

9.3.1   *Submodel 1: Geometry of the submodel*

A model of the reservoir formation is used as a submodel for the well #1 calculation, as shown in Figure 9.6. Its thickness is 0.5 m, diameter is 7 m, and borehole diameter is 8.5 in. (0.2159 m). The drilling process is simulated in the calculation.

9.3.2   *Submodel 1: Boundary condition and loads*

The displacement constraints on all surfaces except the inner borehole surface are derived from the numerical results of the global model at the field scale shown in Fig. 9.1. A set of values of pressure drawdown will be applied on the borehole surface which sets the pore pressure boundary condition at borehole surface.

Gravity and mud weight pressure are applied on the borehole surface along with the initial geostress.

9.3.3   *Numerical scheme of the calculation*

The numerical scheme of the simulation consists of the following five calculation steps:

- Step 1: Establish the initial geostress and boundary conditions and apply the gravity load; perform a 3D FEM calculation with porous flow coupled with deformation.

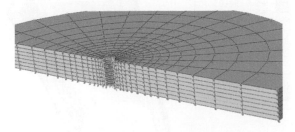

Figure 9.6.   Submodel 1: geometry of the submodel for well #1.

- Step 2: Remove the borehole elements (simulating drilling) and apply mud weight pressure on the borehole surface.
- Step 3: Reduce/replace the mud weight pressure with bottomhole pressure (BHP), which equals the reservoir pressure.
- Step 4: With reference to safe pressure drawdown, further reduce the pressure on the borehole surface as well as the pore pressure at the boundary of the borehole surface.
- Step 5: Check the area of plastic deformation and the value of plastic strain, and make a recommendation for the CVPDD design.

Generally, if there is no plasticity, there will be no sanding potential. If the plastic strain is very small, there will be no obvious sanding. Obvious sanding will occur when the plastic strain is large enough.

9.3.4 *Numerical results*

The following cases of pore pressure boundaries were simulated numerically:

- Initial geostress and pore pressure field.
- Stress and pore pressure field after the wellbore was drilled.
- Status as mud weight was replaced by BHP, which is 20.5 MPa.
- Status as pressure drawdown was set as 344738 Pa (50 psi).
- Status as pressure drawdown was set as 689476 Pa (100 psi) and up to 4826332 Pa (700 psi) with 689472 Pa (100 psi) interval.

Fig. 9.7 through Fig. 9.15 show the following numerical results of variables:

- Distribution of equivalent plastic strain.
- Plastic region where plastic strain occurs.

Table 9.2 summarizes the numerical results of plastic strain corresponding to the set of CVPDDs. The critical value of equivalent plastic strain is chosen at 1%, which means that the value of pressure drawdown that can lead to a plastic strain of more than 1% will cause obvious sanding production.

Based on the numerical results shown in Fig. 9.7 through Fig. 9.15, and those listed in Table 9.2, it is suggested here that CVPDD is 2757904 Pa (400 psi). In practice, careful observation of the sanding phenomena is required after the pressure drawdown exceeds 2068428 Pa (300 psi). It would be very helpful if any in-situ field laboratory can be available for sanding production in this field: parameter calibration can be made accordingly and accuracy will be further ensured then.

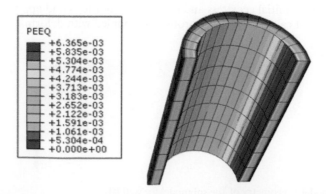

Figure 9.7.   Distribution of equivalent plastic strain when mud weight was replaced by BHP.

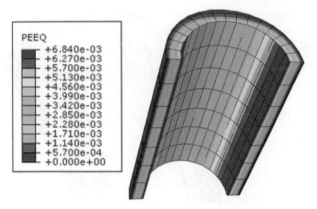

Figure 9.8.   Distribution of equivalent plastic strain when pressure drawdown was set at 344738 Pa (50 psi).

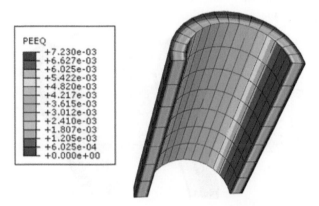

Figure 9.9.   Distribution of equivalent plastic strain when pressure drawdown was set at 689476 Pa (100 psi).

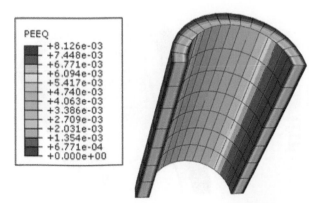

Figure 9.10.   Distribution of equivalent plastic strain when pressure drawdown was set at 1378952 Pa (200 psi).

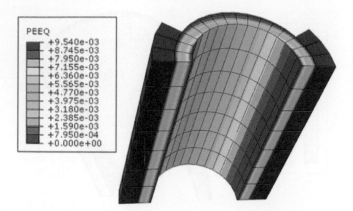

Figure 9.11.   Distribution of equivalent plastic strain when pressure drawdown was set at 2068428 Pa (300 psi).

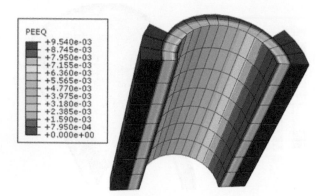

Figure 9.12.   Distribution of equivalent plastic strain when pressure drawdown was set at 2757904 Pa (400 psi).

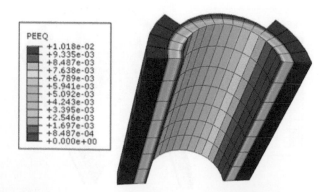

Figure 9.13.   Distribution of equivalent plastic strain when pressure drawdown was set at 3447380 Pa (500 psi).

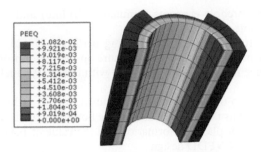

Figure 9.14.   Distribution of equivalent plastic strain when pressure drawdown was set at 4136856 Pa (600 psi).

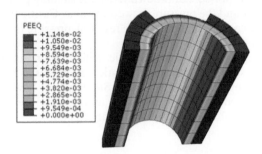

Figure 9.15.   Distribution of equivalent plastic strain when pressure drawdown was set at 4826332 Pa (700 psi).

Table 9.2.   Summary of plastic strain corresponding to the set of CVPDDs.

Pressure Drawdown (PDD)/Pa	Maximum equivalent plastic strain (Peeq)/%	Notice
BHP (0 drawdown)	0.64	
344738	0.68	
689476	0.72	
1378952	0.83	
2068428	0.89	Critical value of equivalent plastic strain is chosen as 1%.
2757904	0.95	
3447380	1.02	
4136856	1.08	
4826332	1.15	

9.4   CASE 2: NUMERICAL PREDICTION OF CVPDD FOR WELL WITH CASING COMPLETION

The procedure for predicting the CVPDD of a cased well is similar to that used for an open-hole completion, but with an additional step performed to simulate the perforation tunnel, as follows:

- Step 1: Establish the initial geostress and boundary conditions and apply gravity load; make a 3D FEM calculation with porous flow coupled with deformation.

- Step 2: Remove the borehole elements (simulating drilling) and apply the mud weight pressure on the borehole surface.
- Step 3: Remove the tunnel elements (simulating perforating) and apply the initial pore pressure and surface pressure/mud weight pressure on the perforation tunnel surface.
- Step 4: Reduce/replace the mud weight pressure with the BHP, which equals the reservoir pressure.
- Step 5: With reference to the safe pressure drawdown, further reduce the pressure on the borehole surface as well as the pore pressure at the boundary of the borehole surface.
- Step 6: Check the area for plastic deformation and the value of plastic strain, and make a recommendation for the CVPDD design.

An additional variable for the analysis of CVPDD with a casing completion is the number of shots per ft alongside the pressure drawdown value; the sanding potential depends on both the number of shots per ft and the pressure drawdown value. Casing failure is another factor to be considered in some cases as an unfavorable result of pressure drawdown; however, it is not considered here for brevity.

Generally, the perforation will create plastic deformation around the perforation tunnel. The total plastic strain is the sum of plastic strain caused by the perforation and the plastic strain caused by pressure drawdown. Sanding will have less influence on the wellbore stability for a cased well section than for an openhole well section. Thus, casing completion allows a larger value of plastic strain than that allowed for a well with openhole completion.

9.4.1   *Modeling casing*

Simplifications made in the modeling of casing is an important aspect in this calculation. The function of casing in completion is to enhance the stability of the wellbore. A concrete ring between the casing and formation seals the oil/gas within the formation. In the production process, both the concrete ring and the casing are nonpermeable and have no influence on the porous flow occurring within the formation. Therefore, to reduce the computational burden, the details of the casing and concrete ring can be neglected. However, their nonpermeable property and boundary effect is essential in the model.

The membrane elements used here are to simulate the casing function. Its stiffness is rather low, and the thickness is only 1 mm. Zero radial displacement constraints are applied to all nodes of the membrane elements. Low stiffness values will reduce the shear stress in the formation connected to the membrane. In this way, the major mechanical properties of the physical phenomena have been ensured with the least computational burden.

The model of the membrane is built to the model by using 'offset mesh' technique of Abaqus CAE. Key sentences in the data of the model are listed below. Only one line of element nodal information is provided in this list. Metric units have been adopted in the expression of mechanical variables and parameters.

```
*Element, type=M3D8R
5585, 14, 270, 2032, 313, 8126, 8203, 8204, 8205
... ...
*Membrane Section, elset=OffsetElements-1, material=LOWSTIFF, controls=EC-1
0.001,
*Material, name=LOWSTIFF
*Elastic
10., 0.2
```

9.4.2 *Case 2A: Casing with perforation of 8 shots per 0.3048 m*

9.4.2.1 *Description of the model: Case 2A*

In this example, a well section with a perforation density of 8 shots per 0.3048 m (per ft) is chosen as Case 2A for the CVPDD calculation.

A model of reservoir formation with the following values of geometric parameters is adopted.

The thickness of the model is set at 0.1524 m (0.5 ft), with four shots in the model (i.e., 8 shots per 0.3048 m (or say per ft)). The diameter of the model is 7 m, and the diameter of borehole is 8.5 in. (0.2159 m). The diameter of the perforation tunnel is set at 1 in. (0.0254 m).

The displacement constraints on all surfaces except the inner borehole surface and perforation tunnel surfaces are derived from the numerical results of the global model at the field scale, as shown in Fig. 9.1.

On the inner surface of the borehole, the membrane layer is used to simulate the impermeable property of the casing. Perforation tunnels punch through the impermeable membrane.

A set of pressure drawdown values will be applied on the surface of the perforation tunnels. This sets the pore pressure boundary condition at surfaces of the perforation tunnels.

Gravity load and initial stress are applied to the entire model, and pressures are applied on the surface of perforation tunnels corresponding to various values of pressure drawdown.

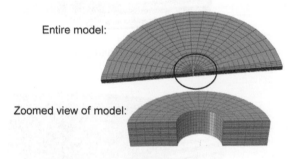

Figure 9.16. Geometry of the model: Case 2A (8 shots per 0.3048 m).

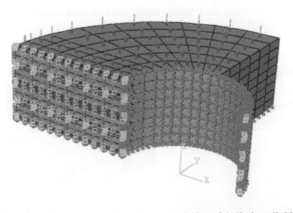

Figure 9.17. Boundary conditions and loads of the model: Case 2A (8 shots/0.3048 m).

9.4.2.2   *Numerical results of Case 2A*

The following cases of pore pressure boundaries have been simulated numerically:

- Initial geostress and pore pressure field.
- Stress and pore pressure field after the wellbore is drilled.
- Status as mud weight was replaced by BHP, which is 38 MPa.
- Status as pressure drawdown was set at 3447380 Pa (500 psi).
- Status as pressure drawdown was set at 4136856 Pa (600 psi).

Fig. 9.18 and Fig. 9.19 show the numerical results for the distribution of equivalent plastic strain around the perforation tunnel.

9.4.3   *Case 2B: Casing with perforation of 4 shots per 0.348 m (per ft)*

9.4.3.1   *Geometry of the model: Case 2B*

The values of the parameters of the submodel for Case 2B are almost the same as those used in Case 2A, except that the number of shots per 0.3048 m (per ft) of the perforation tunnel is 4 rather than 8 as previously used (Fig. 9.20).

Figure 9.18.   Distribution of equivalent plastic strain around perforation tunnel with drawdown at 3447380 Pa (500 psi) for Case 2A.

Figure 9.19.   Distribution of equivalent plastic strain around perforation tunnel with drawdown at 4136856 Pa (600 psi) for Case 2A.

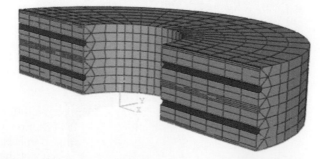

Figure 9.20.   Model of Case 2B: with 2 shots in 0.1524 m (1/2 ft) of thickness.

9.4.3.2 *Numerical results of Case 2B*

The following cases of pore pressure boundaries have been simulated numerically:

- Initial geostress and pore pressure field.
- Stress and pore pressure field after wellbore is drilled.
- Status as mud weight was replaced by BHP, which is 38 MPa.
- Status as pressure drawdown was set at 3447380 Pa (500 psi).
- Status as pressure drawdown was set at 4136856 Pa (600 psi).

Fig. 9.21 through Fig. 9.23 show the numerical results for the distribution of equivalent plastic strain around the perforation tunnel for the situations with 0 Pa, 3447380 Pa (500 psi), and 4136856 Pa (600 psi) pressure drawdown, respectively.

Fig. 9.21 shows that the perforation process could result in equivalent plastic strain of as much as 0.392 around the perforation tunnel. However, it exists only around the tunnel and does not propagate far away. Conversely, the perforation process is a squeezing process, which should cause stiffening of the sand around the tunnel hole. Therefore, the influence of this plastic strain on the stability of the tunnel hole is negligible.

Fig. 9.22 shows that the maximum plastic strain after a 3447380 Pa (500 psi) pressure drawdown is 0.4075. However, the increment resulting from this pressure drawdown is only:

$$0.4075–0.392 = 0.0149.$$

The amount, 0.392, is caused by the perforation process.

Fig. 9.23 shows that the maximum plastic strain after 4136856 Pa (600 psi) pressure drawdown is 0.4128. However, the increment resulting from this pressure drawdown is only:

$$0.4128–0.392 = 0.0202.$$

The 100 psi increase in pressure drawdown results in a plastic strain increase of 0.0053 which is the difference between 0.0202 and 0.0149.

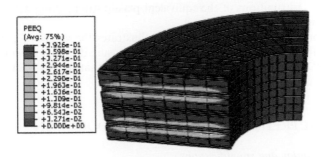

Figure 9.21. Distribution of equivalent plastic strain around perforation tunnel with drawdown at 0 Pa for Case 2B.

Figure 9.22. Distribution of equivalent plastic strain around perforation tunnel with drawdown at 3447380 Pa (500 psi) for Case 2B.

Figure 9.23.   The distribution of equivalent plastic strain around the perforation tunnel with drawdown at 4136856 Pa (600 psi) for Case 2B.

After considering all of the previously mentioned factors, the 4136856 Pa (600 psi) pressure drawdown is concluded to be safe for oil production without obvious sand production.

9.4.4   *Remarks*

With the fully coupled, poro-elastoplastic finite element method, 3D numerical analyses were performed to predict the CVPDD for wells with openhole completion and wells with casing completion. A submodeling technique was adopted to address the discrepancy between the scale of the oil field and that of the wellbore section. A global model at the field scale was used to establish the geostress field of the wells. The pressure drawdown and poro-elastoplastic behavior of the formation near the wellbore was calculated with the submodel, which has a diameter of 7 m.

Based on the results provided by the 3D finite-element calculation, the critical pressure drawdown before the onset of sanding was predicted, and a CVPDD of 2757904 Pa (400 psi) was suggested for the well with openhole completion and 4136856 Pa (600 psi) for well with casing completion. Distributions of the equivalent plastic strain, along with a plastic region where plastic strain occurs, were illustrated.

The visualization of the 3D numerical results illustrates the values of plastic strain and shows the size of the plastic region under the given pressure drawdown, which are useful in selecting the CVPDD. The results presented here indicate that the 3D FEM is a highly efficient tool for predicting CVPDD.

9.5   NUMERICAL PREDICTION OF SANDING PRODUCTION

9.5.1   *Model description and simplifications*

The numerical simulation of sand production is an important component of the application of the multiphysics phenomena in the petroleum industry. In the process of sand production, a total of three phases of material (liquid, solid matrix, and solid grain) are involved. During sand production, sand grain is carried to the borehole by oil and transported to the surface. The sand can be transported by oil to the borehole both from the formation interior and exterior. As a consequence of sanding, the perforation tunnel will become ovalized by the erosion on tunnel surface.

In Abaqus, the sanding is simplified as surface erosion, and this process is simulated by the adaptive meshing technology (Dassault Systems 2008b). In the calculation of erosion/ sanding, the given adaptive mesh domain will be re-meshed by using the same mesh topology which accounts for the new location of tunnel surface. Consequently, the enlargement of the perforation tunnel as a result of sand production is simulated.

The adaptive meshing is treated as a post-converged increment process. Fig. 9.24 shows the numerical scheme for the calculation of adaptive re-meshing.

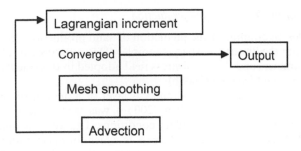

Figure 9.24. The remeshing process.

The numerical scheme for the calculation of erosion consists of the following:

- Define the domain at which the adaptive re-meshing calculation will be performed. Erosion will occur on the external surface of the adaptive domain.
- The erosion equation defines the movement of the external surface where erosion occurs. Its mathematical expression will be introduced into the Abaqus finite element calculation through the user subroutine UMESHMOTION. This subroutine will use the numerical solutions of nodal variables, such as pore pressure, and elemental material properties, such as porosity, to calculate the velocity of the erosion speed.
- Calculate the amount of sand production by integrating the volume change within the domain of adaptive meshing. The specific variable VOLC is designed to calculate the amount of sanding production.

The erosion equation adopted in this calculation was proposed by Papamichos and Malmanger (2001) and is expressed as:

$$V_e = \lambda(1-n)cv_w \qquad (9.1)$$

where V_e represents the erosion velocity, v_w is the pore fluid velocity, c is the transport concentration, n is the porosity, and λ is the sand production coefficient. The value of λ depends on the equivalent plastic strain $\bar{\varepsilon}^{pl}$. The value of λ is set to be zero when $\bar{\varepsilon}^{pl}$ is less than a given cutoff value $\bar{\varepsilon}^{pl}_{cutoff}$. It is also limited by a model constant λ_2. The expression of λ in terms of equivalent plastic strain is:

$$\lambda = \lambda_1\left(\bar{\varepsilon}^{pl} - \bar{\varepsilon}^{pl}_{cutoff}\right) \quad \text{if } \bar{\varepsilon}^{pl} \geq \bar{\varepsilon}^{pl}_{cutoff}; \quad \lambda_1 \text{ is a model constant;} \qquad (9.2)$$

Users can implement their own mathematical expression in the MESHMOTION subroutine if they have their own equation for erosion velocity that differs from that expressed in Eq. 9.1.

9.5.2 *Numerical procedure for prediction of sand production*

The numerical procedure for the prediction of sand production can be briefly stated as follows:

- Perform a geostatic calculation to build the geostress field of the model.
- Model the change, removing the wellbore and then setting casing.
- Apply the boundary conditions of the displacement constraints and pore pressure, as well as the initial pore pressure conditions.
- Set the pressure drawdown to perform a steady state analysis.
- Perform a consolidation analysis with the material erosion for a time period to simulate the production process.
- Use VOLC to calculate the amount of sand production.

9.5.3   *An example of prediction of sand production*

This section provides an example of the prediction of sand production; the geometry of this example is taken from the library of Abaqus samples (Dassault Systems 2008b). The values of the material parameters and the initial stress field adopted in this calculation are different from one given in Dassault Systems (2008b).

Fig. 9.25 shows the geometry of the example. A layer of a quarter of circle with a thickness of 0.203 m (8 in). and a radius of 5.08 m (200 in). was used in the model because of the symmetry of the domain and of its boundary conditions. The wellbore and one perforation tunnel are included. This model simulates the case in which four perforation tunnels are uniformly distributed in a horizontal circle. The diameter of the wellbore is 0.3175 m (12.5 in)., and the diameter of the perforation tunnel is 0.0432 m (1.7 in).

Fig. 9.26 shows the domain in which adaptive meshing will be performed during the calculation. It includes the entire area where the perforation tunnel is located. Fig. 9.27 shows the cut-view of the geometry of the perforation tunnel.

The initial geostress field is set as:

$$\sigma_x = -35.2 \text{ MPa}; \sigma_y = -34.5 \text{ MPa}; \sigma_z = -51.7 \text{ MPa}; \tau_{xy} = \tau_{yz} = \tau_{xz} = 0$$

The initial pore pressure is 41.37 MPa (6,000 psi); uniform pressure is applied to the top of the model, which is P = 93.08 MPa (13,500 psi).

The values of the material parameters are set as follows: porosity = 0.25; Young's modulus E = 9.1 GPa, Poisson's ratio = 0.22; permeability = 325 Darcy, with day as the time scale.

The Drucker-Prager criterion is adopted for the plastic loading judgment. The values of parameters are set as: $d = 6.2$ MPa, $\beta_{\text{friction}} = 40°$ and $\beta_{\text{dilatancy}} = 35°$.

With the numerical procedure shown in the previous subsection, the deformed mesh and shape of perforation tunnel after sanding/erosion is shown in Fig. 9.28. An enlargement

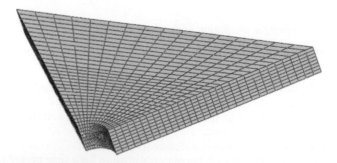

Figure 9.25.   The model used in the prediction of sand production.

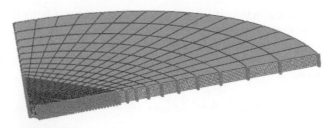

Figure 9.26.   The adaptive zone in the model.

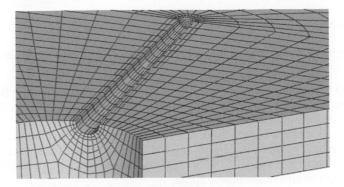

Figure 9.27.   The initial shape of the perforation tunnel.

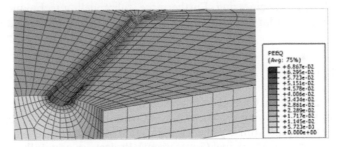

Figure 9.28.   Final shape of the perforation tunnel after erosion/sanding and the distribution of equivalent plastic strain around it.

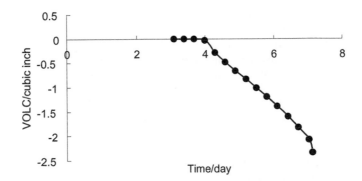

Figure 9.29.   The volume of sand production in a four-day interval (ranges from the beginning of the fourth day to the end of the seventh day).

occurs to the perforation tunnel in the section near the wellbore. The distribution of the equivalent plastic strain is also shown in Fig. 9.28; the maximum value reaches 6.867%.

Fig. 9.29 shows the amount of sand production within the four-day period. This is the amount of sand produced by this perforation tunnel only. The amount of sand production corresponds to the erosion of the formation and to the volume reduction of the matrix; thus, its value is set as negative in the calculation. Fig. 9.29 shows that the predicted amount of sand production stabilized after a four-day period, with no further significant increase.

9.6   CONCLUSIONS

From the numerical results presented in this chapter, the following conclusions can be derived:

- The CVPDD for openhole completions can be effectively predicted with 3D FEM. The criterion based on the critical value of equivalent plastic strain along with the size of the plastic domain is reasonable to be used for the selection of the CVPDD.
- For wellbores with casing completions, it is necessary to distinguish between the amount of equivalent plastic strain caused by drilling and perforation and the amount that resulted from pressure drawdown; the latter should be used in the selection of the CVPDD, rather the total amount of equivalent plastic strain.
- Numerical results have validated that the 3D FEM can effectively predict the amount of sand production with a given condition of oil production, such as pressure drawdown and the property of material strength. However, parameter calibration is important for a practical calculation in engineering. Furthermore, submodeling techniques are useful in providing accurate boundary conditions for the model used to predict sand production.

ACKNOWLEDGEMENTS

Partial financial support from China National Natural Science Foundation through contract 10872134, support from Liaoning Provincial Government through contract RC2008-125 and 2008RC38, support from the Ministry of State Education of China through contract 208027, and support from Shenyang Municipal Government's Project are gratefully acknowledged.

NOMENCLATURE

E	=	Young's modulus, F/L^2, Pa
v	=	Poisson's ratio
σ_x	=	Effective stress component in x-direction, Pa
σ_y	=	Effective stress component in y-direction, Pa
τ_{xy}	=	Effective shear stress component in xy-plane, Pa
τ_{xz}	=	Effective shear stress component in xz-plane, Pa
τ_{zy}	=	Effective shear stress component in zy-plane, Pa
σ_z	=	Effective vertical stress component, Pa
ρ	=	Density, kg/m^3
V_e	=	Erosion velocity, m/s
v_w	=	Pore fluid velocity, m/s
λ	=	Sand production coefficient
$\bar{\varepsilon}^{pl}_{cutoff}$	=	Given cutoff value of the equivalent plastic strain
$\bar{\varepsilon}^{pl}$	=	Equivalent plastic strain
c	=	Transport concentration
d	=	Parameter in Drucker-Prager model as cohesive strength, Pa
m	=	Model parameter in creep model
n	=	Porosity
β	=	Parameter in Drucker-Prager model as friction angle, $^\circ$
3D FEM	=	3-dimensional finite element method
BHP	=	Bottomhole pressure, Pa
CVPDD	=	Critical value of pressure drawdown, Pa
TVD	=	True vertical depth, m

REFERENCES

Dassault Systems: *Abaqus analysis user's manual*. Vol. 3: Materials, Version 6.8, Vélizy-Villacoublay, France: 19.3.1-17–19.3.2-14, 2008a.

Dassault Systems: *Abaqus example problems manual*. Vol. 1: Static And Dynamics Analysis, Version 6.8, Vélizy-Villacoublay, France: 1.1.22-1–1.1.22.11, 2008b.

He, G.J.: *Investigation on fundamentals of increasing heavy oil production with limited sand production for unconsolidated sand reservoir* (in Chinese). PhD Thesis, Southwest University of Petroleum, Chengdu, 2005.

Liu, X.X.: *Mathematical modeling on coupled three-phase problems and its finite element method formulation* (in Chinese). PhD Thesis, Southwest University of Petroleum, Chengdu, 2005.

Oluyemi, G.F. and Oyemeyin, M.B.: Analytical critical drawdown (CDD) failure model for real-time sanding potential prediction based on Hoek and Brown failure criterion. *J. Petrol.Gas Engin.* 1:2 (2010), pp. 16–27.

Papamichos, E. and Malmanger, E.M.: A sand-erosion model for volumetric sand predictions in a North Sea reservoir. SPE 69841, *Reservoir Evaluation & Engineering* 45 (2001) pp. 44–50.

Papamichos, E. and Stavropoulou, M.: An erosion-mechanical model for sand production rate prediction. *Int. J. Rock Mech. Mining Sci.* 35 (1998) pp. 531–532.

Shen, X.: Subsidence prediction and casing integrity with respect to pore-pressure depletion with 3-D finite element method. Paper SPE 138338 presented at the SPE Latin American & Caribbean Petroleum Engineering Conference, Lima, Peru, 1–3 December, 2010.

van den Hoek, P.J., Hertogh, G.M.M., Kooijman, A.P., de Bree, Ph., Kenter, C.J. and Papamichos, E: A new concept of sand production prediction: theory and laboratory experiments. Paper SPE 36418 presented at the SPE Annual Technical Conference and Exhibition, Denver, CO, USA, 6–9 October 1996.

Wang, Y. and Lu, B.: A coupled reservoir-geomechanics model and applications to wellbore stability and sand prediction. Paper SPE 69718 presented at the SPE International Thermal Operations and Heavy Oil Symposium, Porlamar, Margarita Island, Venezuela, 12–14 March, 2001.

Yi, X., Valko, P.P. and Russell, J.E.: Predicting critical drawdown for the onset of sand production. Paper SPE 86555 presented at the SPE International Symposium and Exhibition on Formation Damage Control, Lafayette, LA, USA, 18–20 February, 2004.

Zhuang, Z., Zhang, F. and Cen, S.: *Examples and analysis of ABAQUS nonlinear finite element method* (in Chinese). Academic Press, Beijing, China, 2005.

CHAPTER 10

Cohesive crack for quasi-brittle fracture and numerical simulation of hydraulic fracture

Xinpu Shen

10.1 INTRODUCTION

As a major measure of reservoir stimulation, hydraulic fracturing has been investigated since the 1950s (Christianovich and Zheltov 1955; Cleary 1980). As a result of the rapid development of unconventional petroleum resources in recent years, the investigation of hydraulic fracturing has attracted the interest of many researchers (Soliman *et al.,* 2004; Ehlig-Economides *et al.,* 2006; Bahrami and Mortazavi 2008; Bagherian *et al.,* 2010).

This chapter introduces the concepts and the numerical processes of modeling hydraulic fractures using the 3-dimensional finite element method (3D FEM). The subsequent sections include the following:

- Cohesive crack model concepts.
- Popular cohesive crack models.
- Cohesive element with pore pressure designed for the simulation of hydraulic fractures.
- Example numerical analysis of the hydraulic fracture of a reservoir formation performed with the Abaqus 3D FEM program.

10.2 COHESIVE CRACK FOR QUASI-BRITTLE MATERIALS

10.2.1 *Concepts of cohesive crack*

Generally speaking, the quasi-brittle fracture is a major fracture type that describes the crack initialization and propagation within rock-like materials, such as rock and concrete. Many researchers have used various quasi-brittle models in recent years in their analyses of the fracture of rock-like materials and structures (Hillorberg *et al.,* 1976). As reported by Bazant and Cedolin (1991), Cocchetti (1998), and Cocchetti *et al.,* (2001), the existing constitutive models suitable for the numerical analysis of quasi-brittle fracture can be roughly classified into three groups: mode-I cohesive crack models, mixed-mode cohesive crack models (i.e., Coulomb-type elasto-plastic models), and elasto-plastic constitutive models for classical joint elements that assume a prior existence of discontinuity locus, but non-cohesive process is considered.

With the introduction of cohesive crack models that assume the existence of a fracture process zone ahead of the macro-crack for rock fractures, many studies have focused on the analysis of the fracture of rock and cement-based materials that were subjected to either pure tension or to mixed-mode fracture loading. These models can be used to simulate the fracture process of rock formations; they can also be adopted in the simulation of fractures at the interface of two dissimilar materials and in the simulation of joint behavior related to either masonry or grouted joints. Cohesive crack models for the simulation of the fracture process in rock-like materials consist of the following two fundamental features:

- Linear-elastic material behavior is assumed throughout the solid or structure considered except at the locus of the potential crack.

- Displacement discontinuity is allowed over that locus and is related to traction across it by a suitable relationship. Bazant and Cedolin (1991) and Karihaloo (1995) provide a review (updated to 1991) of abundant literature concerning the cohesive crack model.

The following vectors and relationships define the crack surface of a cohesive crack, as shown in Fig. 10.1:

$$\mathbf{p} = \mathbf{p}^- = \mathbf{p}^+, \quad \mathbf{w} = \mathbf{u}^+ - \mathbf{u}^- \qquad (10.1)$$

$$\mathbf{p} = \begin{Bmatrix} p_n \\ p_t \end{Bmatrix}, \quad \mathbf{w} = \begin{Bmatrix} w_n \\ w_t \end{Bmatrix} \qquad (10.2)$$

where $\mathbf{p}^-$ is the traction vector calculated from the lower part of the solid, $\mathbf{p}^+$ is the traction vector calculated from the upper part of the solid by FEM, and $\mathbf{p}$ is the traction vector act on the interface. $\mathbf{w}$ is the vector of displacement discontinuity across the interface crack; subscripts "n" and "t" denote the components in the normal direction and tangential direction, respectively. For the traction calculated with FEM, a positive sign is used for the traction components in the positive axial direction, and a negative sign is used for those opposite to its related axis. For interface traction components, a positive sign is used for tensile, and a negative sign is used for compression. The FEM formulated with generalized and interface variable provides the fundamental relationship expressed in Eq. 10.1.

The sign conventional is defined for the interface traction in the way that positive represent tensile and negative for compression. However, for the traction calculated from the FEM, the positive sign is for the one in the positive direction of the global coordinates. Therefore, the sign of the interface variable should be changed when using the cohesive crack variables in a global FEM description.

10.2.2   *Influence of hydraulic pressure on yielding conditions*

While the hydraulic pressure in the process zone being accounted for, Terzaghi's assumption is adopted:

$$p_n = p'_n - p_n^{(f)} \qquad p_t = p'_t \qquad (10.3)$$

where p_n is effective stress of the skeleton and p'_n is the total stress. Water pressure in the process zone, i.e., $p_n^{(f)}$, does not influence the distribution of the traction component in the tangential direction.

According to the linear-assumption of the water pressure distribution inside a cohesive zone, as shown in Fig. 10.2, there is the following expression of $p_n^{(f)}$ for the mixed-mode crack:

$$p_n^{(f)} = \frac{w_n - \lambda}{w_{nc}} \hat{p}_n \qquad (10.4)$$

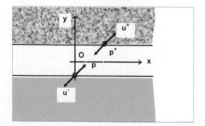

Figure 10.1.   Interface variables.

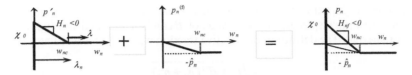

Figure 10.2.   The influence of hydraulic pressure on the cohesive crack behavior.

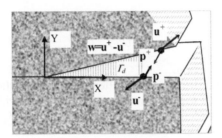

Figure 10.3.   Illustration of 3D interface variables of mixed-mode crack.

where $\hat{p}_n$ is the value of the hydraulic pressure at the mouth of the process zone, i.e., the tip of the macro-crack; w_{nc} is the critical value of the normal crack opening (also known as the crack opening displacement (COD) in fracture mechanics), and w_n is the normal displacement at a point in the interface process zone. λ is the plastic multiplier for the inelastic crack opening in the normal direction of the crack surface, and its function is to maintain the value of $p_n^{(f)}$ at a point in the process zone to be not larger than $\hat{p}_n$.

10.2.3   *Cohesive models for mixed-mode fracture*

Fig. 10.3 shows the interface variables for a mixed-mode cohesive fracture; the definitions of these variables include the following: $\mathbf{p}$ represents the traction vector on the interface, and $\mathbf{p} = -\mathbf{p}^+ = \mathbf{p}^-$, $\mathbf{u}^+$ and $\mathbf{u}^-$ represent the displacement vector on two sides of the interface, respectively; $\mathbf{w}$ represents the vector of the crack opening, i.e., the displacement discontinuity across the interface. The definition of traction vector and the displacement vector for mixed-mode crack are shown in Eq. 10.5.

$$\mathbf{p} = \begin{Bmatrix} p_n \\ \mathbf{p}_t \end{Bmatrix}, \quad \mathbf{w} = \begin{Bmatrix} w_n \\ \mathbf{w}_t \end{Bmatrix} \tag{10.5}$$

where p_n is the normal traction component, $\mathbf{p}_t$ is the tangential traction vector, w_n is the normal component of the crack opening, $\mathbf{w}_t$ is the tangential vector of the crack opening. In the problems considered in this chapter, the deformations are assumed to be small in the sense that the equilibrium relations are not influenced by the configuration changes. It is also assumed that the mechanical process is an isothermal process.

10.2.4   *Cohesive model of effective opening for mixed-mode crack*

To build the connection between the experimental results obtained by tests of the mode-I crack and mixed-mode fracture phenomena, which is analogous to the von Mises method adopted for equivalent stress in plasticity, Camacho and Ortiz (1996) proposed the concept of effective crack opening. The definition of the effective opening displacement w is:

$$w = \sqrt{w_n^z + \beta^{-z}(w_{t1}^z + w_{t2}^z)} \tag{10.6}$$

where β is the weight parameter through which shear displacements are assigned to the normal crack opening displacement. The value of β can be calibrated through experiments for various materials. The definition of effective traction was also presented as:

$$p = \sqrt{p_n^z + \beta^{-z}(p_{t1}^z + p_{t2}^z)} \tag{10.7}$$

Based on the first and second laws of thermodynamics, the general **p-w** relationship was given in vector form as:

$$\mathbf{p} = \begin{Bmatrix} p_n \\ p_t \end{Bmatrix} = \frac{\partial \varphi}{\partial \mathbf{w}} \tag{10.8}$$

where φ is the free energy function.

For different materials and structures, the softening law for interface effective variables can be determined on the basis of experimental results. Fig. 10.4 provides one type of softening law between the effective crack opening and the effective traction. Camacho and Ortiz (1996) provide the mathematical description:

$$p = e p_n^u \frac{w}{w_c} e^{-\frac{w}{w_c}} \tag{10.9}$$

where e is the Nepero's number 2.71828, and w_c is a characteristic opening displacement in correspondence of p_n^u. If unloading is considered for the holonomic model case shown in Fig. 10.4 (a), the unloading path will follow the reverse direction of the loading path. For the nonholonomic case shown in Fig. 10.4 (b), the maximum crack opening displacement achieved in the loading process, w_{max}, should be recorded for the calculation of the unloading process. This model accounts only for the degradation of elastic stiffness, as shown in Fig. 10.4 (b); it does not account for residual displacement for unloading. Then, the unloading law can be written as described by Camacho and Ortiz (1996):

$$p = \frac{p_{max}}{w_{max}} w, \quad \text{if} \quad w < w_{max} \text{ or } \dot{w} < \tag{10.10}$$

The effective opening model has had numerous applications in the quasi-brittle fracture analysis for mixed-mode crack in recent years. Results reported in Camacho and Ortiz (1996) show that this model successfully simulates the mixed-mode fracture phenomenon. A problem with this model lies in the choice of the value for parameter β: no rigorous theory has been presented for this issue. Generally speaking, this model is an effective vehicle for mixed-mode

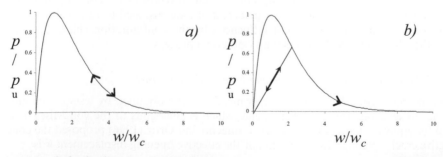

Figure 10.4.   Softening laws in terms of effective opening: a) holonomic model; b) nonholonomic.

fractures as long as mode-I opening, rather than friction sliding, is the dominant mechanism in the fracture process. If friction sliding is the dominant mechanism in the fracture process, this model is not a good choice.

This model can be used for the dynamic analyses of dams under the joint action of earthquake loading and working water pressure.

10.2.5   *Cohesive law formulated in standard dissipative system*

As an alternative tool to describe the elastoplastic interface cohesive model with Coulomb type models, Corigliano (1993) and Bolzon and Corigliano (1997) presented a model in the standard dissipative system for the cohesive crack model. The interface of thickness t is assumed to have existed prior to the appearance of any crack, and it is the loci of crack propagation. The general description of an elastoplastic constitutive relationship for interface variables is presented as follows (Corigliano 1993; Bolzon and Corigliano 1997):

$$
\begin{gathered}
\mathbf{w} = \mathbf{w}^e + \mathbf{w}^p \\
\mathbf{p} = \frac{\partial \Psi\left(\mathbf{w}^e, \boldsymbol{\alpha}\right)}{\partial \mathbf{w}^e}, \quad \mathbf{a} = \frac{\partial \Psi\left(\mathbf{w}^e, \boldsymbol{\alpha}\right)}{\partial \boldsymbol{\alpha}} \\
\dot{\mathbf{w}}^p = \dot{\lambda}\frac{\partial f^T\left(\mathbf{p}, \mathbf{a}\right)}{\partial \mathbf{p}}, \quad \dot{\boldsymbol{\alpha}} = \dot{\lambda}\frac{\partial f^T\left(\mathbf{p}, \mathbf{a}\right)}{\partial \mathbf{a}}
\end{gathered}
\tag{10.11}
$$

where $\boldsymbol{\alpha}$ denotes the vector of internal variables, $\mathbf{a}$ denotes the vector of conjugate force corresponding to $\boldsymbol{\alpha}$; Ψ is the free energy function for a point on the cohesive interface. The expression of free energy function is given by Corigliano (1993) as:

$$
\Psi = \frac{1}{2}\left[K_n\left(w_n^e\right)^2 + K_{t1}\left(w_{t1}^e\right)^2 + K_{t2}\left(w_{t2}^e\right)^2\right] + \Psi_2(\boldsymbol{\alpha})
\tag{10.12}
$$

where K_x denotes elastic stiffness in the direction x; $x = n, t1, t2$, $\Psi_2(\boldsymbol{\alpha})$ is an additional function to set the shape of the softening law in $\mathbf{p}$-$\mathbf{w}$ space. Given $\boldsymbol{\alpha}$ as a scalar, there is:

$$
\Psi_2(\alpha) = \alpha + \frac{1}{2}h\alpha^2
\tag{10.13}
$$

where h is a positive non-dimensional parameter, then a nonlinear softening branch can be obtained, as shown in Fig. 10.5.

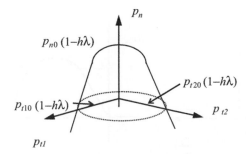

Figure 10.5.   Initial elastic domain.

10.2.5.1   *Elastoplastic damage interface model*

When the elastoplastic damage model was adopted, the activation function of inelastic processes on the cohesive interface was expressed by Corigliano (1993) and Bolzon and Corigliano (1997) as:

$$F = \sqrt{a_n\left(\frac{\langle p_n\rangle_+}{1-d_n}\right)^2 + a_{t1}\left(\frac{p_{t1}}{1-d_{t1}}\right)^2 + a_{t2}\left(\frac{p_{t2}}{1-d_{t2}}\right)^2} - 1 + h\lambda$$

$$a_i = \frac{1}{\left(p_{ic}\right)^2}, \quad i = n, t1, t2$$

(10.14)

where $\langle\bullet\rangle_+$ denotes that only the positive case of the variable in the bracket is accounted for; d_n, d_{t1}, d_{t2} are damage variables, and p_{ic} denotes the critical traction value at which the softening begins to occur in that ith direction. The degradation of stiffness K_i caused by damage was considered in this model, and the free energy for the damaged interface point is:

$$\Psi = \frac{1}{2}(1-d_n)K_n\langle w_n\rangle_+^2 + \frac{1}{2}K_n\langle w_n\rangle_-^2 + \frac{1}{2}(1-d_{t1})K_{t1}w_{t1}^2 + \frac{1}{2}(1-d_{t2})K_{t2}w_{t2}^2 \quad (10.15)$$

where subscript '+' denotes the tensional state and '–' denotes the compressive state. In this model, a unified flow surface is adopted for all irreversible processes, such as damage and plasticity. The evolution laws for damage and plasticity are given as:

$$\dot{\mathbf{w}}^p = \frac{\partial F}{\partial \mathbf{p}}\overset{\circ}{\lambda}, \quad \overset{\circ}{\lambda} = \sqrt{\left(\dot{\mathbf{w}}^p\right)^T\cdot\left(\dot{\mathbf{w}}^p\right)}, \quad \mathbf{d} = \mathbf{L}(h\lambda) = \begin{Bmatrix} 1-\sqrt{1-2\gamma_n h\lambda} \\ 1-\sqrt{1-2\gamma_{t1} h\lambda} \\ 1-\sqrt{1-2\gamma_{t2} h\lambda} \end{Bmatrix}$$

(10.16)

$$F \le 0, F\overset{\circ}{\lambda} = 0, \overset{\circ}{\lambda} \ge 0$$

where $\gamma_i, i = n, t1, t2$ and h are model parameters and can be determined by fracture energy through the following equations:

$$G_{\mathrm{I}} = \frac{1}{K_n a_n}\left(\frac{1}{2} + \frac{1}{15\gamma_n^2} + \frac{1}{3\gamma_n}\right) + \frac{1}{8\gamma_n^2 h} + \frac{1}{2\gamma_n h}$$

$$G_{\mathrm{II}} = \frac{1}{K_t a_t}\left(\frac{1}{2} + \frac{1}{15\gamma_t^2} + \frac{1}{3\gamma_t}\right) + \frac{1}{8\gamma_t^2 h} + \frac{1}{2\gamma_t h}$$

(10.17)

where γ_t denotes both γ_{t1} and γ_{t2}.

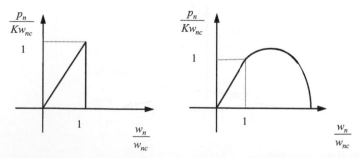

Figure 10.6.   Softening laws: (a) brittle failure (b) progressive failure.

Experimentally speaking, this method (unified flow rule for damage and plasticity) provides a reasonable approximation of physical reality, but it is not a general method for dealing with plural dissipative processes.

10.2.5.2　*Viscoplastic interface crack model*

The fundamental equations of Corigliano's viscoplastic interface constitutive model (Corigliano and Allix 2000) include the following:

$$\frac{\dot{W}^{vp} = \gamma\langle f(\mathbf{p},\lambda)\rangle_{+}^{N}\,\partial f(\mathbf{p},\lambda)}{\partial \mathbf{p}}$$

$$\mathbf{p} = \mathbf{K}\mathbf{w}^{e},\ \mathbf{K} = \mathrm{diag}(K_{i}),\ i = n, t1, t;$$

$$f(\mathbf{p},\lambda) = \sqrt{a_{n}\langle p_{n}\rangle^{2} + a_{t1}p_{t1}^{2} + a_{t2}p_{t2}^{2}} - 1 + h\lambda, \tag{10.18}$$

$$\lambda = \int_{0}^{\tau}\sqrt{(\dot{\mathbf{W}}^{vp})^{T}(\dot{\mathbf{W}}^{vp})}\,\mathrm{d}\tau$$

where τ denotes time. This model follows the exponential visco law proposed by Perzyna (1966). The parameters used in this model include the following: K_{i}, a_{i}, $i = n$, $t1$, $t2$ and h, γ, N.

The interface models formulated in the standard dissipative system successfully simulates the softening behavior and other inelastic properties of the interface under mixed-mode loading.

10.3　COHESIVE ELEMENT COUPLED WITH PORE PRESSURE FOR SIMULATION OF HYDRAULIC FRACTURE OF ROCK

The details of the description of the cohesive element coupled with pore pressure for rock fracture can be found in Dassault Systems (2010). This section introduces some of the features of this type of cohesive element.

10.3.1　*Nodal sequence and stress components of cohesive element*

As shown in Fig. 10.7, the Coh3D8P element is the cohesive element coupled with pore pressure designed to simulate rock fracturing under hydraulic fracture. Fig. 10.8 shows the number sequence of the cohesive element Coh3D8P. Although it is labeled as Coh3D8P, this type of element actually consists of 12 nodes. The nodes on the mid-surface are specially designed for fluid leakoff calculations, and are generated by specifying an '*offset*' action in Abaqus CAE, which will automatically add four mid-nodes to the element in terms of corner nodes.

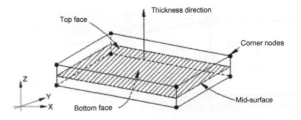

Figure 10.7.　Thickness direction of the cohesive element Coh3D8P.

The stress components designed for this type of cohesive element include the following:

- S33: Direct through-thickness stress
- S13: Transverse shear stress
- S23: Transverse shear stress

The tensors of the mechanical variables of the same type of cohesive element, such as strain, all have the same number of components. The displacement and pore pressure are given as nodal variables in the calculation of this type of cohesive element.

10.3.2   *Fluid flow model of the cohesive element*

In the applications of the cohesive element in hydraulic fracturing, the fluid flow continuity within the gap and through the interface must be maintained to enable the fluid pressure on the cohesive element surface to contribute to its mechanical behavior, which enables the modeling of the hydraulically driven fracture. It also enables the modeling of an additional resistance layer on the surface of the cohesive element.

10.3.2.1   *Defining pore fluid flow properties*

The fluid constitutive response comprises two parts: (1) tangential flow within the gap, which can be modeled with either a Newtonian or power law model; and (2) normal flow across the gap, which can reflect resistance as a result of caking or fouling effects. Fig. 10.9 shows the flow patterns of the pore fluid in the element.

10.3.2.2   *Tangential flow*

By default, there is no tangential flow of pore fluid within the cohesive element. To enable a tangential flow, a gap flow property must be defined in conjunction with the pore fluid material definition in the model input data. There are two types of tangential flow that can be adopted in the calculation: 1) Newtonian fluid and 2) power law fluid. Their expressions are given in the following subsections.

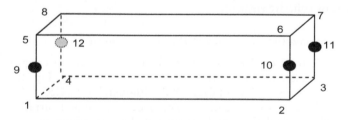

Figure 10.8.   Number sequence of the cohesive element Coh3D8P. Nodes on the mid-surface are specially designed for fluid leakoff calculations.

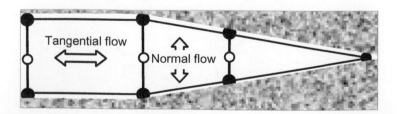

Figure 10.9.   Flow within cohesive elements.

10.3.2.3 *Newtonian fluid*

The volume flow rate density vector **q** of a Newtonian fluid is given by the expression:

$$\mathbf{q}d = -k_t \nabla p \tag{10.19}$$

where k_t is the tangential permeability (the resistance to the fluid flow), ∇p is the pressure gradient along the cohesive element, and d is the gap opening.

In Abaqus software, the gap opening, d, is defined as:

$$d = t_{curr} - t_{orig} + g_{init} \tag{10.20}$$

where t_{curr} and t_{orig} are the current and original cohesive element geometrical thicknesses, respectively; g_{init} is the initial gap opening, which has a default value of 0.002 m.

With reference to Reynold's equation, the tangential permeability is given as:

$$k_t = \frac{d^3}{12\mu} \tag{10.21}$$

where μ is the fluid viscosity.

10.3.2.4 *Power law fluid*

For a power law fluid, its constitutive relation is defined as:

$$\tau = K \dot{\gamma}^\alpha \tag{10.22}$$

where τ is the shear stress, $\dot{\gamma}$ is the shear strain rate, K is the fluid consistency, and α is the power law coefficient. Its tangential volume flow rate density is defined as:

$$\mathbf{q}d = -\left(\frac{2\alpha}{1+2\alpha}\right)\left(\frac{1}{K}\right)^{\frac{1}{\alpha}}\left(\frac{d}{2}\right)^{\frac{1+2\alpha}{\alpha}} \|\nabla p\|^{\frac{1-\alpha}{\alpha}} \nabla p \tag{10.23}$$

10.3.2.5 *Normal flow across gap surfaces*

Normal flow is defined through a fluid leakoff coefficient for the pore fluid material. This coefficient defines a pressure-flow relationship between the middle nodes of the cohesive element and their adjacent surface nodes. The fluid leakoff coefficients can be interpreted as the permeability of a finite layer of material on the cohesive element surfaces.

With the leakoff coefficients c_t and c_b, the normal flow is defined as:

$$q_t = c_t(p_i - p_t) \tag{10.24}$$

and

$$q_b = c_b(p_i - p_b) \tag{10.25}$$

where q_t and q_b are the flow rates into the top and bottom surfaces, respectively. p_i is the midface pressure, and p_t and p_b are the pore pressures on the top and bottom surfaces, respectively.

The following output variables are available when flow is enabled in pore pressure cohesive elements:

- GFVR: Gap fluid volume rate
- PFOPEN: Fracture opening
- LEAKVRT: Leakoff flow rate at element top

- ALEAKVRT: Accumulated leakoff flow rate at element top
- LEAKVRB: Leakoff flow rate at element bottom
- ALEAKVRB: Accumulated leakoff flow rate at element bottom

10.4   NUMERICAL SIMULATION OF HYDRAULIC FRACTURING WITH 3-DIMENSIONAL FINITE ELEMENT METHOD

10.4.1   *Numerical procedure for the numerical simulation of hydraulic fracturing*

The numerical procedure for the numerical simulation of hydraulic fracturing with 3D FEM includes the following:

- Build an initial geostress field, and set the initial values of the initial gap.
- Simulate the drilling of the wellbore with the mud weight pressure applied on the wellbore surface.
- Simulate the casing step by applying new boundary conditions on the wellbore surface.
- Perform the first step of the consolidation analysis, which simulates fluid injection into the formation under pressure. In this process, the cohesive element will open as a result of the fluid pressure and create a hydraulic fracture.
- Perform the second step of the consolidation analysis to enable the pore pressure to dissipate. In this process, hydraulically created fractures are prevented from closure by additional boundary conditions to simulate the effects arising from the proppant material.
- Perform the third step of the consolidation analysis with a pressure drawdown applied to the wellbore nodes. The petroleum production from the formation will be simulated in this step.

This procedure simulates the entire process of a reservoir stimulation and production.

10.4.2   *Finite element model*

This example simulates hydraulic fracturing within the target formation at a true vertical depth (TVD) of 500 m.

The geometry of the casing and cement ring has been simplified in this example: displacement constraints and pore pressure boundary conditions are applied to the inner surface of wellbore, and no casing and cement ring are included in the model. In this way, the function of supporting the casing and cement ring are ensured with less computational burden in the model than would be in a model that includes those details.

10.4.2.1   *Geometry and mesh*

As shown in Fig. 10.10, a model of a semi-circular layer has been adopted for the geometry of the formations investigated. The thickness of the layer is 50 m, with a 20 m target formation in the center. The thickness of the top of the formation is 10 m, and the thickness of the bottom of the formation is 20 m. The radius of the model is 200 m.

The locus of the crack generated by hydraulic pressure is set at the central surface of the semicircle. A set of cohesive elements that have pore pressure as the primary nodal variables were used to simulate the fracture.

10.4.2.2   *Initial conditions*

The initial conditions defined for the model include the following:

- Void ratio: 0.2 for upper shale formation, 0.25 for lower shale formation, and 0.35 for reservoir formation.
- Initial pore pressure: 24.5 MPa for all formations. Simplifications were made for this initial condition.

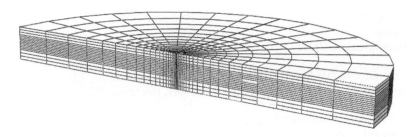

Figure 10.10.   Geometry of the model.

- Initial geostress components: for all formations, they are given the values as $\sigma_x = -11.6$ MPa, $\sigma_y = -12.6$ MPa, $\sigma_z = -16.6$ MPa, and all shear components are zero. This is assumed to correspond to a TVD of approximately 2000 m. These values of geostress are given in the space of effective stress, which is the amount of the total overburden minus the pore pressure. The total overburden stress is 41.1 MPa, which is the sum of the vertical effective stress and the pore pressure.

A value of 2 mm as the initial gap has been assigned on those nodes at where the crack begins under the given hydraulic pressure.

10.4.2.3   *Boundary condition*
Displacement constraints were applied to the normal directions of the outer surface, symmetry surface, and bottom and top surfaces. The pore pressure boundary conditions for the symmetry surface are given as zero. For other boundary surfaces, the pore pressure values are assumed constant and are the same values as that used for the initial pore pressure.

10.4.2.4   *Loads*
The loads applied to this model include the following:

- Geostress.
- Concentrated flow during the injection step.
- Mud weight pressure applied to the inner surface of the wellbore during the drilling step, which is removed after the casing installation.

10.4.2.5   *Values of material parameter*
Because of the complexity of the nonlinear property of the materials, as well as the number of types of different materials, the definition of the material parameters is rather complicated. The data that defines the various properties of the model are provided in the following without details.

```
** MATERIALS
**
*Material, name=COH-Upper
*Damage Initiation, criterion=QUADS, omega=0.1, tolerance=0.2
3.5e+06, 2e+07, 2e+07
*Damage Evolution, type=ENERGY, mixed mode behavior=BK, power=2.284
28000.,280000.,280000.
*Density
2100.,
*Elastic, type = TRACTION
8.5e+10, 8.5e+10, 8.5e+10
```

```
*Fluid Leakoff
1e-08, 1e-08
*Gap Flow, kmax=0.00052
1e-06,
*Material, name=COH-lower
*Damage Initiation, criterion=QUADS, omega=0.1, tolerance=0.2
3.5e+06, 2e+07, 2e+07
*Damage Evolution, type=ENERGY, mixed mode behavior=BK, power=2.284
28000.,280000.,280000.
*Density
2100.,
*Elastic, type=TRACTION
8.5e+10, 8.5e+10, 8.5e+10
*Fluid Leakoff
1e-08, 1e-08
*Gap Flow, kmax=0.00052
1e-06,
*Material, name=COH-middle
*Damage Initiation, criterion=QUADS, omega=0.1, tolerance=0.2
3.5e+06, 2e+07, 2e+07
*Damage Evolution, type=ENERGY, mixed mode behavior=BK, power=2.284
28000.,280000.,280000.
*Density
2100.,
*Elastic, type=TRACTION
8.5e+10, 8.5e+10, 8.5e+10
*Fluid Leakoff
1e-08, 1e-08
*Gap Flow, kmax=0.00052
1e-06,
*Material, name=RESERVOIR
*Elastic
2.3e+10, 0.2
*DENSITY
2100.0
*DAMPING, BETA=0.00323
*CONCRETE DAMAGED PLASTICITY
36.31
*CONCRETE COMPRESSION HARDENING
13.0E+6, 0.0
24.1E+6, 0.001
*CONCRETE TENSION STIFFENING
2.9E+6        ,0
1.94393E+6    ,0.000199
1.30305E+6    ,0.000369
0.873463E+6   ,0.00052
0.5855E+6     ,0.000661
0.392472E+6   ,0.000794
0.263082E+6   ,0.000924
0.176349E+6   ,0.001053
0.11821E+6    ,0.001182
0.0792388E+6  ,0.001313
0.0531154E+6  ,0.00145
```

```
*CONCRETE TENSION DAMAGE
0      ,0
0.381217   ,0.000199
0.617107   ,0.000369
0.763072   ,0.00052
0.853393   ,0.000661
0.909282   ,0.000794
0.943865   ,0.000924
0.965265   ,0.001053
0.978506   ,0.001182
0.9867   ,0.001313
0.99177   ,0.00145
*Permeability, specific=1e-05
1e-10, 0.0001
2e-10, 0.6
*Material, name=ROCK
*Density
2300.,
*Elastic
1.5e+10, 0.27
*Permeability, specific=0.0001
1e-15, 0.001
2e-15, 0.6
*Material, name=UpperShale
*Density
2500.,
*Elastic
1.5e+10, 0.27
*Permeability, specific=0.0001
1e-15, 0.001
2e-15, 0.6
*Material, name=lowerShale
*Density
2500.,
*Elastic
1.5e+10, 0.27
*Permeability, specific=0.0001
1e-15, 0.001
2e-15, 0.6
```

10.4.3   *Numerical results*

The purpose of the calculation is to illustrate and validate the numerical model and the numerical procedure previously described. The calculation required only 350 seconds with the following breakdown: the pumping simulation required 50 seconds; the hold stage required 100 seconds, and the drawdown stage required 200 seconds. In engineering, these processes require 10 times longer than was used in the illustrative calculation. The drawdown period is actually the production period, which can last for months.

Fig. 10.11 through Fig. 10.13 show the numerical results obtained with the model. Fig. 10.11 shows that the fracture was generated within the target formation and began to grow.

Fig. 10.12 shows the variation of the fluid flow rate during the pump down, hold pressure, and drawdown stages.

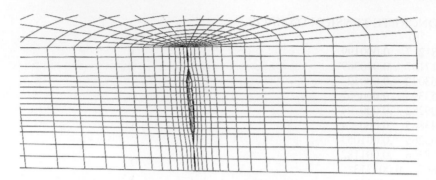

Figure 10.11.   Visualization of numerical solution of crack opening after 400 seconds of hydraulic fracture (100 times enlargement applied to nodal displacement visualization).

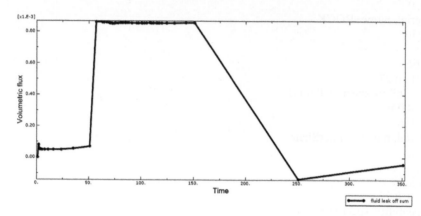

Figure 10.12.   Variation of the fluid flow rate during pump down, hold pressure, and drawdown stages (Unit: m³/s).

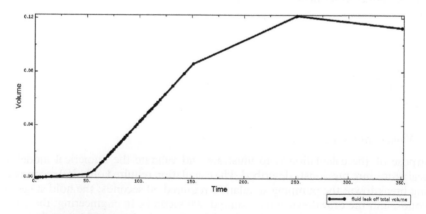

Figure 10.13.   Variation of total volume of fluid flow during the pump down, hold pressure, and drawdown stages (Unit: m³).

Fig. 10.13 shows the variation of the total volume of fluid flow during the pump down, hold pressure, and drawdown stages. The fluid flow volume value is assigned a negative sign for the production/drawdown stage. Fig. 10.13 shows that the total volume of fluid flow declines between 250 and 350 s, which corresponds to a production stage.

10.5 CONCLUSIONS

Cohesive crack for the simulation of quasi-brittle fracturing of rock-like material has been used in engineering for a long time. The cohesive element coupled with pore pressure is a numerical tool that has been widely used in the simulation of hydraulic fracturing in the petroleum industry. This chapter presents primary studies and applications performed by the author.

To improve the accuracy of the numerical results obtained by using the model described in this chapter, the following areas should be the subject of future studies:

- Submodeling techniques should be used in the determination of the initial geostress field. With the global model at a field scale, the submodeling technique can produce accurate boundary conditions for the model at the formation thickness scale.
- The continuum damage model should be included in the model for the simulation of hydraulic fracturing. There is currently only one locus along which the crack can develop. In fact, with the continuum damage model embedded in Abaqus program, it is easy to simulate the propagation of a secondary crack, which is usually perpendicular to the principal crack modeled by the cohesive crack.

ACKNOWLEDGEMENTS

Partial financial support from National Natural Science Foundation of China through contract 10872134, support from Liaoning Provincial Government through contract RC2008-125 and 2008RC38, support from the Ministry of State Education of China through contract 208027, and support from Shenyang Municipal Government's Project are gratefully acknowledged.

NOMENCLATURE

$\mathbf{p}^-$	=	Traction vector calculated from the lower part of the solid, Pa
$\mathbf{p}^+$	=	Traction vector calculated from the upper part of the solid, Pa
$\mathbf{p}$	=	Traction vector act on the interface, Pa
$\mathbf{u}^+$	=	Displacement vector, m
$\mathbf{u}^-$	=	Displacement vector, m
$\mathbf{w}$	=	Vector of displacement discontinuity across the interface crack, m
p_n	=	Effective stress of the skeleton for mode-I crack, and normal traction component for mixed mode crack, Pa
p_n'	=	Total stress, Pa
$p_n^{(f)}$	=	Water pressure in the process zone, Pa
$\hat{p}_n$	=	Value of the hydraulic pressure at the mouth of the process zone, Pa
$\mathbf{p}_t$	=	Tangential traction vector, Pa
$\mathbf{w}_t$	=	Tangential vector of the crack opening, m
w_{nc}	=	Critical value of the normal crack opening, Pa
w_n	=	Normal displacement at a point in the interface process zone, m
λ	=	Plastic multiplier
β	=	Weight parameter
φ	=	Free energy function
e	=	Nepero's number 2.71828
w_c	=	Characteristic opening displacement in correspondence of p_n^u, m
$\boldsymbol{\alpha}$	=	Vector of internal variables
$\mathbf{a}$	=	Vector of conjugate force corresponding to α
Ψ	=	Free energy function
K_x	=	Elastic stiffness in the direction x; $x = n, t1, t2$

$\langle \bullet \rangle_+$	=	Only the positive case of the variable in the bracket is accounted for
d	=	Damage variable
p_{ic}	=	Critical traction value at which the softening begins to occur in that ith direction, Pa
γ_i	=	Model parameter
τ	=	Time, s
$\mathbf{q}$	=	Volume flow rate density vector, m³/s
d	=	Gap opening, m
k_t	=	Tangential permeability, D
∇p	=	Pressure gradient, Pa/m
t_{curr}	=	Current cohesive element geometrical thicknesses, m
t_{orig}	=	Original cohesive element geometrical thicknesses, m
g_{init}	=	Initial gap opening, m
μ	=	Fluid viscosity, Pa·S
$\dot{\gamma}$	=	Shear strain rate
K	=	Fluid consistency
α	=	Power law coefficient
c_t	=	Leakoff coefficient
c_b	=	Leakoff coefficient
q_t	=	Flow rates into the top surface
q_b	=	Flow rates into the bottom surface
p_t	=	Pore pressures on the top surface
p_b	=	Pore pressures on the bottom surface
COD	=	Crack opening displacement, m
FEM	=	Finite element method

REFERENCES

Bagherian, B., Sarmadivaleh, M., Ghalambor, A., Nabipour, A., Rasouli, V. and Mahmoudi, M.: Optimization of multiple-fractured horizontal tight gas well. Paper SPE 127899-MS presented at the SPE International Symposium and Exhibition on Formation Damage Control, Lafayette, LA, USA, 10–12 February 2010.

Bahrami, V. and Mortazavi, A.: A numerical investigation of hydraulic fracturing process in oil reservoirs using non-linear fracture mechanics. Paper ARMS5-2008-120 presented at the 5th Asian Rock Mechanics Symposium, Tehran, Iran, 24–26 November, 2008.

Bazant, Z.P. and Cedolin, L.: *Stability of structures: elastic, inelastic, fracture and damage theories*, Oxford University Press, Oxford, New York, 1991.

Bolzon, G. and Corigliano, A.: A discrete formulation for elastic solids with damaging interfaces. *Comp. Methods Appl. Mech. Engng.* 140:2 (1997), pp. 329–359.

Camacho, G.T. and Ortiz, M.: Computational modeling of impact damage in brittle materials. *Int. J. Solids Struct.* 33:20–22 (1996), pp. 2899–2938.

Christianovich, S.A. and Zheltov, Y.P.: Formation of vertical fractures by means of a highly viscous fluid. *Proc. 4th World Petroleum Congress*, 2 (1955), pp. 579–586.

Cleary, M.P.: Comprehensive design formulae for hydraulic fracturing. Paper SPE 9259-MS presented at the SPE Annual Technical Conference and Exhibition, Dallas, TX, USA, 21–24 September, 1980.

Cocchetti, G.: *Failure analysis of quasi-brittle and poroplastic structures with particular reference to gravity dams*. PhD Thesis, Politecnico di Milano, Italy, 1998.

Cocchetti, G., Maier, G. and Shen, X. P.: On piecewise linear models of interfaces and mixed mode cohesive cracks, 2002, Vol. 3, no. 3: 279–298.

Corigliano, A.: Formulation, identification and use of interface models in the numerical analysis of composite delamination. *Int. J. Solids Struct.*, 30:20 (1993), pp. 2779–2811.

Corigliano, A. and Allix, O.: Some aspects of interlaminar degradation in composites. *Comput. Methods Appl. Mech. Engng.* 185 (2000), pp. 203–224.

Dassault Systems: Abaqus analysis user's manual. Vol. 3: Materials, Version 6.8, Vélizy-Villacoublay, France: 19.3.1-17–19.3.2-14, 2008.

Ehlig-Economides, C.A., Valko, P. and Dyashev, I.: Pressure transient and production data analysis for hydraulic fracture treatment evaluation. Paper SPE 101832 presented at the 2006 SPE Russian Oil and Gas Technical Conference and Exhibition, Moscow, Russia, 3–6 October, 2006.

Hillorberg, A., Modeer, M. and Petersson, P.E.: Analysis of crack formation and crack growth in concrete by means of fracture mechanics and finite elements. *Cement Concrete Res.* 6 (1976), pp. 773–782.

Karihaloo, B.L.: *Fracture mechanics and structural concrete.* Longman Scientific & Technical, Harlow, Great Britain, 1995.

Perzyna, P.: Fundamental problems in visco-plasticity. *Adv. Appl. Mech. 2* (1966), pp. 343–377.

Soliman, M.Y., East, L. and Adams, D.: Geo-mechanics aspects of multiple fracturing of horizontal and vertical wells. Paper SPE 86992-MS presented at the SPE International Thermal Operations and Heavy Oil Symposium and Western Regional Meeting, Bakersfield, CA, USA, 16–18 March, 2004.

CHAPTER 11

Special applications in formation stimulation and injection modeling

Mao Bai

11.1 INTRODUCTION

To avoid the pitfalls that may be associated with using only rules of thumb in hydraulic fracturing design, modeling becomes an indispensable component. This chapter describes special numerical applications in drilling and completion that can supplement useful guidelines in the experience-based design. Focusing on the practical side of applications for formation stimulation and injection using numerical modeling, the discussion is restricted to the introduction of certain special issues associated with various unconventional numerical applications without presenting the detailed procedure and results.

Numerical modeling can be used as a predictive tool during the initial stages of fracturing design and as an interpretative tool when field observations become available to rationalize behavior and to determine whether or not alternate actions are required. In addition, modeling can be used to validate design parameters and to verify the post-stimulation (injection) effect to improve the control of the overall stimulation (injection) process. Although realistic models follow fundamental laws (e.g., momentum, mass, and energy conservations) and constitutive relationships, sensible guidance from modeling usually still requires the proper calibration of the initial and boundary conditions from field feedback and measurements. Modeling is commonly integrated with the subsequent design, operations, and monitoring to provide assurance for successful implementation.

Predictive modeling is also used to assist in making operational decisions, such as determining the following:

- Total injection volume and injection rate for maximized fracturing efficiency.
- Injection fluid and its viscosity for achieving desirable fracture geometries and proppant placement.
- Pumping schedule and proppant concentration for optimized propped volume and fracture conductivity.

The quality of the modeling depends on the quality of the input data. For quality assurance purposes, the input data should be developed from experience, laboratory testing, and field measurements, and the data must be validated against field observations. For hydraulic fracturing, the critical input parameters to be validated consist of the following:

- In-situ stress in relation to the fracture containment.
- Fracture toughness or an equivalent parameter in relation to the fracture propagation.
- Fluid loss in relation to the leakoff volume.
- Fluid rheology in relation to the fracture geometry.
- Solids concentration in relation to formation plugging, screenout, and leakoff.

The ultimate purposes of the numerical model are to optimize the operations with the calculated predictions and assessments. Rather than discussing the modeling techniques for conventional (or normal) model applications in the operations (e.g., hydraulic fracturing in brittle sandstone formations), this chapter focuses on introducing the special unconventional applications for the fracturing modeling in a brief fashion, which includes the stimulation and injection in the following four areas:

- Unconventional tight shale gas reservoirs for reservoir stimulation in completion.
- Cuttings re-injection for solid waste disposal in drilling.
- Frac-pack for stimulating weak sandstone reservoirs in completion.
- Produced water re-injection for liquid waste disposal in production.

11.2   NORMAL APPLICATIONS

For the low permeability formations, the flow rate toward the wellbore can be restricted to such a state that the formation natural conductivity must be stimulated to achieve more efficient production. The artificial fracture around the wellbore perforations can be hydraulically created from the combination of the following conditions:

- Using a faster fluid pumping rate into the wellbore than the fluid leaking rate into the formation.
- Increasing the injection fluid pressure to a magnitude that exceeds the formation strength to generate a planar fracture around wellbore (The fracture orientation is dictated by the in-situ stress configurations, with the direction of the largest fracture dimension being parallel to the orientation of the maximum principal stress when the fracture is not contained in any particular direction.)
- Maintaining the injection pressure or flow rate to continuously enlarge the planar fracture. After the injection ceases, the generated hydraulic fracture will be kept open by injecting proppant (e.g., reinforced sand, such as resin-coated sand) of a certain concentration.

The hydraulic fracture stimulates the hydrocarbon production from the well by the following:

- Bypassing the near-wellbore formation damage zone formed during drilling and completion.
- Extending the conductive flow path deep into the formation (depending on the fracture geometry).
- Modifying the hydrocarbon flow toward the wellbore from the low rate of interstitial pore fluid flow to the high rate of channeled fracture flow.

Therefore, the traditional hydraulic fracturing is characterized by the following requirements:

- A relatively lower permeability formation to minimize the leakoff (Fracturing in high permeability formation is non-traditional; it requires using special techniques, such as tip screenout and frac-pack, to maximize the fracture width while restricting the fracture growth in length and height.)
- A relatively higher injection rate to create sufficiently higher injection pressure to breakdown the formation.
- Continuous injection with the relatively smaller volume because the fracture geometry is frequently dictated by the formation stiffness, and the size of the injected volume becomes less influential.

Hydraulic fracturing has been a primary method for reservoir stimulation and production enhancement. Because of its complex coupled processes involving the concurrent rock deformation, fluid flow, and solid transport, the simulation of hydraulic fracturing in the

Table 11.1.   Advantages and limitations of numerical models for hydraulic fracturing simulation.

Model	Advantage	Limitation
Finite difference method	It is simple because only mathematical equations must be discretized.	It requires simple boundary conditions. It may not accurately represent mechanical effects because the physical representation at the element level is not required.
Boundary element method	The solution beyond the fracture boundary is accurately represented.	Internal fracture shape is not evaluated; therefore, the fracture development is not sensibly evaluated.
Finite element method	The solution inside the fracture boundary is accurately represented.	Solutions at the fracture boundary and beyond are only approximately derived. The boundary conditions are difficult to determine.
Hybrid boundary element and finite element method	The solutions beyond the fracture boundary and at internal areas are accurately represented.	It requires more in-depth understanding of complex physical and mathematical processes.

field applications has been dominated by the numerical methods. In the numerical models, four types are generally available for commercial use:

- Finite difference.
- Boundary element.
- Finite element.
- Hybrid boundary-finite element methods.

Table 11.1 lists the advantages and limitations of these four numerical methods. One of the normal applications for the numerical hydraulic fracturing simulation is the reservoir stimulation and production enhancement in a brittle and relatively low permeability formation (e.g., sandstone formation with less than 50 mD or greater than 1 mD of formation permeability).

Generally speaking, the theoretical formulation of a hydraulic fracturing simulation is related to the following procedures:

- Conservation of momentum that is associated with the stress equilibrium, and coupling with injection as well as with the reservoir pore pressure. For most numerical models, a 3D stress field can be established.
- Conservation of mass that is related to the flow continuity in terms of counteracting injection pressure and reservoir pore pressure. The injection pressure must overcome the in-situ stress to create the hydraulic fracture. The fluid flow is usually represented by the 2D model, and the flow along the direction of the fracture width is ignored. The combination of 3D solid stress and 2D fluid pressure leads to a planar 3D fracture geometry generated by the hydraulic fracturing injection.
- Conservation of mass that is associated with mass balance in the form of proppant transport under the driving force of the concentration gradient. The mechanism of particle deposition throughout the proppant placement process dictates the solids distribution and eventual propped fracture conductivity.
- Either the fracture toughness in fracture mechanics or other formation damage mechanism controls the fracture propagation during the injection process.
- The total injected volume is the summation of the propped fracture volume and the leakoff volume.

For the normal stimulation modeling by hydraulic fracturing, readers may refer to the following literature: Economides and Nolte 2000; Hubbert and Willis 1957; Haimson and Fairhurst 1967; Clifton and Abou-Sayed 1981.

11.3   SPECIAL APPLICATIONS

Over the last three decades, the hydraulic fracturing methods have evolved from a stimulation tool designated only for the conventional brittle reservoir with a relatively low permeability to a primary tool for stimulating various types of reservoirs, ranging from ductile shales to unconsolidated sandstones. In addition, the applications are no longer restricted to the reservoir stimulation and production enhancement. One of the significant applications of the unconventional hydraulic fracturing is used in the petroleum waste disposal, which includes but is not limited to cuttings re-injection and produced water re-injection. To differentiate the numerical simulations of "unconventional" applications of hydraulic fracturing from the traditional applications, the numerical applications described in this chapter are considered as special simulation cases.

Past experience indicates that the direct adoption of conventional hydraulic fracturing simulation to the unconventional hydraulic fracturing simulation may not work. Modifications to the procedure using the conventional hydraulic fracturing model must be made to adapt the traditional model to the special simulation environment. The theoretical differences between the conventional hydraulic fracturing model and the unconventional hydraulic fracturing model must be expounded for a better understanding of the key steps in analyzing the special applications.

With respect to the special applications of stimulation and injection modeling briefly discussed in this chapter, additional information can be found from the following sources:

- Unconventional shale gas stimulation: Bai *et al.*, 2005; Halliburton Consulting 2010; Anthony 2010; Sarout *et al.*, 2006.
- Cuttings re-injection: Bai *et al.*, 2010; Bai *et al.*, 2006; Guo and Geehan 2004.
- Frac-pack in stimulating soft sandstone formation: Bai *et al.*, 2003; Abou-Sayed *et al.*, 2004; Khodaverdian and McElfresh 2000.
- Produced water re-injection: Bai *et al.*, 2009; Bachman *et al.*, 2003; Farajzadeh 2004.

11.4   UNCONVENTIONAL SHALE GAS RESERVOIRS

11.4.1   *Theoretical basis in simulation*

As a result of increasing world oil prices, depleting conventional gas reserves, and a severe mismatch between the energy demand and hydrocarbon production, attention has been focused toward enhancing the natural gas production in various types of gas reservoirs (e.g., tight-gas sands, coalbed methane, and gas hydrates), even in tight-shale gas formations that were previously assumed to be unproductive. As a result of exceptionally low permeability in the tight shales, hydraulic fracturing has become a dominant means of modifying the reservoir original permeability to such a level that natural gas can be smoothly produced.

Because of the special characteristics of tight-shale gas formations, the conventional concepts of hydraulic fracturing may become irrelevant; consequently, the new concepts must be developed to optimize the stimulation process for the tight formations.

Because the shale formation permeability is so low (i.e., ranging from less than 0.1 millidarcy to macro darcy, or even to nano darcy in tight shale gas formations), unlimited gas flow to the well is largely expected to come from the natural fractures intersected by the hydraulic fracture. The ability to simulate the interaction between the hydraulic fracture and the conducting natural fractures is a challenge to the current modeling capability, although many of these techniques are currently being developed (e.g., Gu and Weng 2010).

Without explicitly modeling the interactive flow between hydraulic fractures and natural fractures, the amenable approach is to use the equivalent model with respect to the effective flow from the reservoir to the well that considers the equivalent effect of intersected natural fractures by the hydraulic fracture.

11.4.2   *An equivalent shale gas hydraulic fracturing model*

Leakoff has been considered as a key factor that has a critical effect on the fracture geometry. For conventional hydraulic fracturing in the brittle formations, the general leakoff is usually at the lower end, such as between 10 to 50% (Note: leakoff percentage is defined as the ratio of leakoff volume to total injected volume.). It can be deduced that the fracture volume can be quite large for the brittle formations. For hydraulic fracturing in the soft formations (e.g., in unconsolidated sandstones), the leakoff can be very significant, such as in the range between 80 to 99% (Bai *et al.*, 2003). Under this condition, the tip screenout is purposely created to inhibit the growth in the fracture area (i.e., length and height). Consequently, the fracture width is inflated under the continuing injection. For hydraulic fracturing in the tight-shale gas formations (TSG), experience indicates that the leakoff range is between 50 to 80% (Bai *et al.*, 2005). The leakoff is commonly placed in the same category as formation permeability. For the TSG, apparent formation permeability can be extremely small, which can make gas reservoir non-producible. However, the intrinsic formation permeability in the TSG counts the permeabilities from both rock matrix and natural fractures over the reservoir domain, which can, therefore, yield moderate to relatively high leakoff from the TSG (i.e., leakoff percentage of 50 to 80%).

In the case study, the distributions of in-situ stress and formation leakoff areas along the depth are about uniform between pay zone and bounding layers, which indicate that the fracture growth is not restricted to the pay zone only. Other key parameters include the following:

- The reservoir pore pressure gradient at 0.43 psi/ft (1 SG).
- The injection fluid is Newtonian with a viscosity of 14 cP (0.014 Pa s).
- The injection rate is 30 bpm (286.2 m³/h) with a total injected volume of 3,800 bbl (604.15 m³).
- Eight pumping stages are designed with a pad in the first stage, and proppant concentration is ramped from 0.25 to 1 lb/gal (29.96 to 119.83 kg/m³) over the remaining seven stages.
- The proppant mesh size is 40/60.

In this study of leakoff alone, the extreme conditions of leakoff are examined. In other words, the selected leakoff range is beyond the suggested range of 50 to 80% for the TSG, as shown in Table 11.2 for three different leakoff cases.

Fracture boundary profiles for three different leakoff cases are shown in Fig. 11.1. For the high leakoff case, the fracture area is small, roughly circular, and near the initial perforation center. For the moderate leakoff case, the fracture upward growth dominates with the limited length development. For the low leakoff case, the fracture initially grows in the upward direction and later grows horizontally with the significant development of fracture length.

Comparisons of fracture width and proppant-density (ratio of leakoff volume to total injected volume) contours are shown in Fig. 11.2. In comparison, higher leakoff leads to smaller fracture width and smaller fracture area, but greater proppant density.

Table 11.2.   Three different formation leakoff cases.

Case	Leakoff (%)
Low leakoff	20
Medium leakoff (base)	76
High leakoff	94

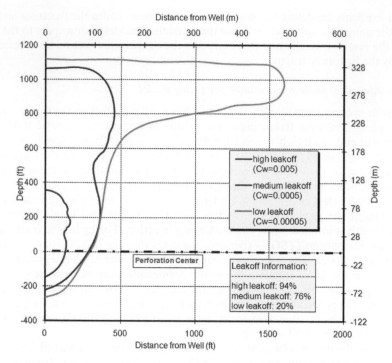

Figure 11.1. Fracture boundary cross-section outlines for three leakoff conditions. The depth is relative to the original perforation center (i.e., 0 depth). The stress is lower at shallower depth (i.e. 8300 psi at relative depth of 1200 ft, i.e., 57.27 MPa at 365.76 m) and is greater at deeper depth (9500 psi at at relative depth of –400 ft, i.e., 65.55 MPa at –121.92 m).

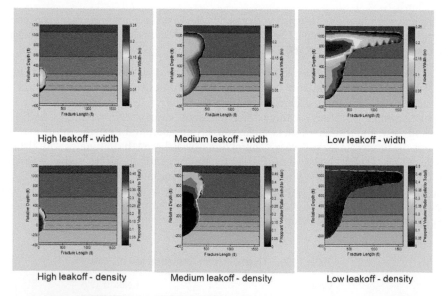

Figure 11.2. Contours of fracture widths and proppant densities for three leakoff cases. The density is defined as the ratio of leakoff volume to the injected volume (1 ft = 0.3048 m; 1 in = 2.54 cm).

Table 11.3.   Propped fracture lengths and widths under different leakoff cases.

Simulation name and leakoff ratio	Leakoff coeff. ($10^{-5} \times$ m/min$^{0.5}$)	Half fracture length (m)	Maximum fracture width (mm)
Leakoff-24%	2.44	506	9.7
Leakoff-38%	4.88	454	9.4
Leakoff-47%	7.32	403	8.6
Leakoff-60%	12.2	332	8.1
Leakoff-69%	18.3	276	7.6
Leakoff-80%	30.5	200	7.1

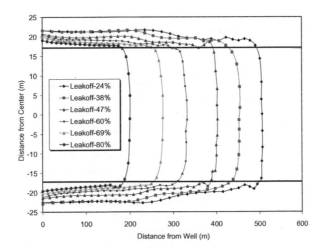

Figure 11.3.   Cross sectional view of fracture geometries under six leakoff cases. All fractures grow slightly out of the pay zone.

11.4.3   *Leakoff effect for a contained fracture*

The case in Section 11.4.2 depicts the fracture geometric evolution when the hydraulic fracture is not contained within the pay zone. This usually occurs when the contrasts of either stresses or leakoffs between the pay zone and bounding layers are small. If stress or leakoff contrasts are relatively significant, the hydraulic fracture is usually contained within the pay zone.

Defining the leakoff percentage as the leakoff volume divided by the total injected volume, Table 11.3 shows the propped fracture half lengths and maximum fracture widths under six leakoff scenarios with the stress contrasts between pay zone and bounding layers. Fig. 11.3 shows the corresponding fracture outlines for the six cases. Generally, higher leakoff results in smaller fracture lengths and smaller fracture widths (Table 11.3). In addition, the fractures are not fully contained within the pay zone for all six cases (Fig. 11.3). The slight fracture height penetrations into the bounding layers reflect the accurate computation using the 3D numerical simulator based on the hybrid finite element and boundary element methods (see Table 11.1).

11.4.4   *Concluding remarks*

For simulating the hydraulic fracturing in unconventional shale gas formations, the existing simulators for the conventional fracturing simulation can be used with some modifications to the input parameters. The most important input parameter is the leakoff coefficient in the form of leakoff percentage. Considering the significant effect of natural fractures on the hydrocarbon migration from the formation to the well, the leakoff percentage between 50 and 80% may be appropriate for simulating the tight shale gas reservoir using an equivalent shale gas model. The effects of the leakoff percentage on both contained fracturing and non-contained fracturing are presented. The propped fracture geometry is a function of the leakoff percentage. The larger the leakoff percentage, the smaller the fracture geometry. The fracture containment is dictated by the stress and permeability contrasts between pay zone and bounding layers. A better containment requires a greater contrast. For the contained fracturing, the fracture may still slightly grow out of pay zone. To prevent this occurrence, the larger leakoff area may be required.

11.5   CUTTINGS RE-INJECTION

11.5.1   *Theoretical basis in simulation*

In traditional hydraulic fracturing (HF), boosting the fracturing efficiency is important because the maximized fracture geometry and minimized leakoff volume can create the optimized hydraulic conductivity for the fluid flow toward the producing wells. In the cuttings re-injection hydraulic fracturing (CRI-HF), maximizing the total cuttings disposal volume is important because the "disposal domain" can be far greater than that defined by the single planar fracture alone. Note that in the petroleum industry, CRI-HF is related to the deep injection where the waste contaminant transport may pose little or no risk to the natural resources, such as aquifers. In other industry and environmental sectors, however, CRI-HF is associated with the shallow injection where the waste contaminant transport must be restricted to certain areas to safeguard the subsurface natural resources. Table 11.4 summarizes the major differences between HF and CRI-HF.

Among all major differences, the difference in injection style is the most important one. In contrast to the continuous injection for the conventional fracturing, the primary purposes of intermittent (or cyclic) injection for the cuttings re-injection (CRI) are to minimize the pressure buildup and detrimental screenout for the prolonged and maximized injection volume while allowing time to prepare the cuttings slurry in a daily operation cycle.

11.5.2   *An equivalent cuttings re-injection model*

Because of the differences between HF and CRI-HF, modifications of the traditional simulator must be made for the CRI applications. The major modification is the pumping schedule because the continuous injection in the HF must be changed to the intermittent (or cyclic) injection in the CRI-HF.

Table 11.4.   Major differences between HF and CRI-HF.

Comparing Item	HF	CRI-HF
Injection rate	Normal to high	Low
Injected volume	Small to moderate	Large
Injection period	Short	Long
Injection style	Continuous	Intermittent
Injecting formation	Non-shale	Including shale
Solid concentration	Moderate to high	Low
Number of fractures	Single	Multiple
Formation leakoff	Small to moderate	Moderate to high
Screenout	OK for frac-pack	Avoided

In the petroleum solid waste disposal industry, the primary purpose of the injection by means of hydraulic fracturing is to inject as much solid waste as possible through the disposal well. Consequently, the efforts are made to prevent any screenout from occurring by maintaining or sometimes reducing, the injection pressure to the acceptable level.

The cyclic injection (various alternative terms used include intermittent injection, batch injection, repeated injection, or periodic injection) is recognized as an effective means of solid waste disposal by injection. The injection involves the daily cycle of injection and shut-in for a designated disposal period. Cyclic injection enables the continuing pressure buildup throughout the injection period to be avoided because the injection pressure will be released during the shut-in time.

Although the total injected volume may be identical between the continuous injection and the cyclic injection, the resulting fracture geometries can be different. Recognizing this difference is important because the zoning of the disposal areas can vary if different injection methods are applied.

As an example of a three-cycle injection (i.e., 16 hours of injection and 8 hours of shut-in daily operation for three days), Fig. 11.4 compares the net pressures; Fig. 11.5 compares the cross-section fracture outlines between continuous injection and cyclic injection.

11.5.3　*Key input parameters for cuttings re-injection modeling*

Because of the numerous differences between HF and CRI-HF (see Table 11.4), many input parameters for the CRI-HF should be uniquely defined with respect to the specific requirements of the CRI. Table 11.5 lists some specific definitions of the key input parameters for the numerical modeling.

Table 11.6 lists examples of some key input parameters. These parameters are case—specific and should be used only as references, rather than as benchmark inputs.

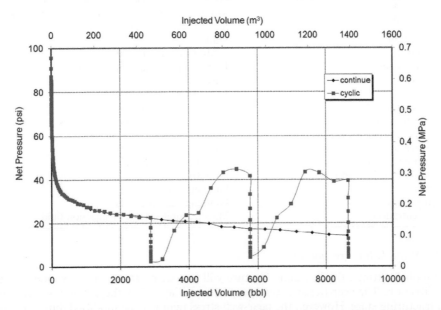

Figure 11.4.　Comparison of net pressures between continuous injection and cyclic injection for a three-day injection event. The net pressures are identical between the two injections for the first injection period. In addition, it appears that the net pressure from the continuous injection is averaged from that of cyclic injection.

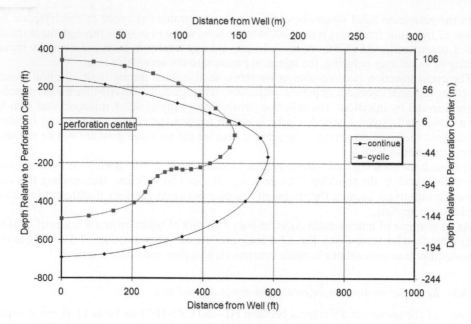

Figure 11.5.   Comparison of cross-section fracture outlines between continuous injection and cyclic injection for a three-day injection event. From the comparison, it appears that the cyclic injection leads to a smaller fracture geometry with an abnormal fracture shape.

Table 11.5.   Definitions of some key input parameters for injection modeling.

Parameter	Note
Stress	Minimum principal horizontal stress for vertical well
Stress gradient	Stress variation within formation layers
Young's modulus	Static
Poisson's ratio	Undrained and drained
Fracture toughness	From linear fracture mechanics
Permeability	Layer property
Porosity	Layer property
Spurt loss coefficient	Layer and injection fluid properties
Leakoff coefficient	Layer and injection fluid properties
Reservoir pore pressure	Linear variation along the depth
Reservoir temperature	Linear variation along the depth
Injection fluid density	From fluid mechanics
Fluid viscosity or rheology constants	From fluid mechanics
Injection rate	Pumping stage property
Injection volume	Pumping stage property
Solid concentration	Mass per unit volume of injection fluid

11.5.4   *Multiple fracture modeling*

In the global settings, the hydraulic fracture will propagate along the maximum principal stress direction. The near-field fracture propagation may follow these global settings at the initial fracturing stage. However, the near-well stress field may be modified during the injection process as a result of the following:

- Poroelastic effect from stress concentration and pressure buildup
- Existence of local heterogeneities, such as natural fractures

Table 11.6. Some key input parameters for an injection modeling case (IC—injection center).

Item	Unit	Value	Unit (SI)	Multiply by
Stress at injection center IC	psi	1,680	MPa	0.0069
Young's modulus at IC	psi	168,600	MPa	0.0069
Poisson's ratio at IC		0.35		
Fracture toughness	psi in$^{0.5}$	1,000	MPa m$^{0.5}$	0.0011
Permeability at IC	mD	695	m^2	9.87×10^{-16}
Porosity at IC	%	10		
Leakoff coefficient at IC	ft/min$^{0.5}$	0.0057	m/min$^{0.5}$	0.3048
Spurt loss at IC	gal/ft^2	0.1	m^3/m^2	0.0004
Reservoir pressure	psi	1,110	MPa	0.0069
Perforation Interval	ft	2,450–2,490	m	0.3048
Injection fluid unit weight	lbf/ft^3	74.9	kg/m^3	16.0185
Fluid rheology constant n'		0.39		
Fluid rheology constant K'	lbf s^n/ft^2	0.0575	kg s^n/m^2	4.8826
Injection rate	bbl/min	2	m^3/min	0.159
Pad volume	bbl	100	m^3	0.159
Rest injection volume	bbl	1,500	m^3	0.159
Solid concentration	lb/gal	0.8	kg/m^3	119.83
Solid particle diameter	in.	0.0165	m	0.0254

- Effect of formation anisotropies from layer modulus and from layer permeability
- Non-symmetric fracture development

Because of the local stress modifications, the hydraulic fracture may reorient during the injection process, leading to the multiple fractures generation.

If the assumption that all disposed solids will be held within the hydraulic fracture is accurate, the ability to model the multiple fractures becomes crucial because the conventional single bi-winged fracture cannot accommodate the disposed solid volume. Because of the complications of simulating multiple fracture development, the rigorous theoretical models for multiple hydraulic fractures are not known to exist. In practice, multiple fracture modeling has been circumvented through invoking the principle of superposition using a number of single fractures. For this reason alone, only the concept of modeling the multiple fractures for the disposal injection will be discussed.

For batch injection, two possibilities exist for the new fracture extension that occurs with the new batch: the previous fracture may continue to extend or a new fracture may form. Fig. 11.6 shows two modes of multiple fractures. In the first mode, labeled as "Multiple fractures 1" in the top of Fig. 11.6 [(a) and (b)], the old fracture of a previous injection is re-opened during the new batch injection and continues to extend, still in a planar fracture manner. The fracture always seeks the path of least resistance to develop because it is easier to extend the old fracture rather than to create a new fracture; consequently, the old fracture will re-open and extend. The extended part of the fracture forms a new fracture because of the difference in the fracture widths between the closed section of the fracture and reopened part of the fracture. "Multiple fractures 1" in Fig. 11.6 shows the progressive old fracture re-opening for each new injection, extending along the initial fracturing orientation in the form of a new fracture.

A second fracture mode is referred to as "Multiple fractures 2," and is shown at the bottom of Fig. 11.6 [(c) and (d)]. This mode is a new fracture extending from the perforated interval of the wellbore at a different direction than that of the previous fracture. After this new fracture has begun, it will continue in a manner similar to that of "Multiple fractures 1" until it is again easier to begin a new fracture at a different orientation than to extend it in the same direction.

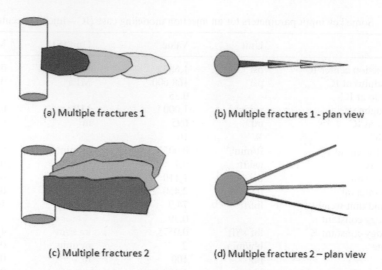

(a) Multiple fractures 1   (b) Multiple fractures 1 - plan view

(c) Multiple fractures 2   (d) Multiple fractures 2 – plan view

Figure 11.6.   Cyclic injection induced multiple fractures: (a) 3D view of Multiple fractures 1; (b) plan view of Multiple fractures 1; (c) 3D view of Multiple fractures 2; (d) plan view of Multiple fractures 2. As the injection continues, Multiple fractures 1 extends laterally (fracture open and re-open mode along the same orientation); Multiple fractures 2 rotates along the vertical axis of the well (beginning new fracture at a different orientation).

Because the general fractures maintain the planar shapes, the simulation of the mode of "Multiple fractures 1" can be accomplished by using an equivalent cuttings re-injection hydraulic fracturing simulator, as shown in Section 11.5.2 (i.e., by cyclic injection method). The remaining questions are to determine how many injection cycles can be simulated under the mode of "Multiple fractures 1" and when a new fracture will begin from a different orientation under the mode of "Multiple fractures 2."

11.5.5   *Net pressure responses in cyclic injection*

Net pressure responses provide the signatures to interpret the following important events from the hydraulic fracturing operation:

- Fracture geometric development.
- Tip screenout.
- Proppant placement.

The pressure responses can be calibrated by the micro-seismic measurements.

Fig. 11.7 compares the net pressures between the continuous injection and cyclic injection for a free fracture extension in the mode of "Multiple fractures 1" shown in Section 11.5.4. From the third cycle onward, the net pressure peaks for the cyclic injection become increasingly greater than the pressure for the continuous injection.

In contrast to the free fracture extension in the "Multiple fractures 1" mode shown in Fig. 11.7, Fig. 11.8 provides the net pressure responses of a restrict fracture growth in the "Multiple fractures 1" mode in which the comparison of net pressures between continuous injection and cyclic injection is given. Because of the differences in the fluid viscosities and injected solids concentrations, the scenario shown in Fig. 11.8 may not directly relate to the scenario shown in Fig. 11.7. However, Fig. 11.8 shows a rising trend of net pressure, as opposed to the declining trend shown in Fig. 11.7. In Fig. 11.8 for the cyclic injection, apparent screenout seems to have occurred during the sixth injection stage. During this final stage after a rapid increase in net pressure, the net pressure suddenly decreases, implying the re-growth of the existing fracture after the screenout. Alternatively, the screenout during the

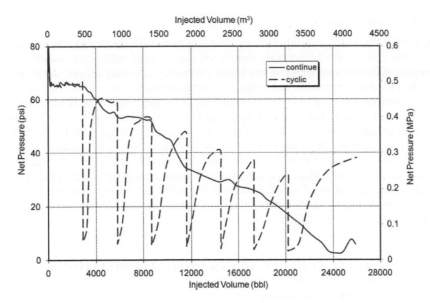

Figure 11.7. Comparison of net pressure responses between continuous injection and cyclic injection for an eight-day injection event. From the comparison, it appears that the cyclic injection leads to an uneven pressure decline than that of continuous injection even though both trends of decline are similar.

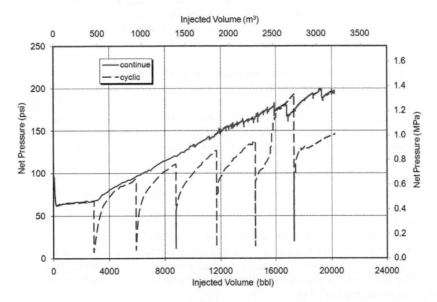

Figure 11.8. Comparison of net pressure responses between continuous injection and cyclic injection for a seven-day injection event. From the comparison, it appears that the cyclic injection leads to a generally smaller pressure than continuous injection with an exception at the sixth cycle when the screenout occurs.

sixth injection cycle may lead to the generation of another fracture at a different orientation at the perforations as shown for the "Multiple fractures 2" mode in Section 11.5.4. Another observation from Fig. 11.8 is that the net pressure for the cyclic injection is generally smaller than that of the continuous injection, except when the screenout occurs at the sixth cycle for the cyclic injection.

11.5.6   *Concluding remarks*

For simulating the cuttings re-injection by means of hydraulic fracturing, the existing simulators for the conventional fracturing simulation can be used with some modifications on the pumping schedule and on some key input parameters. Among others, the most important modification of a conventional hydraulic fracturing simulator is to change the injection mode from continuous type to the cyclic type. The cuttings re-injection model is built based on this feature. In addition to the change in the injection style, many input parameters for injection modeling must change to maximize the injected cuttings volume. In reality, the injected cuttings cannot be contained with a single planar shape hydraulic fracture. As a result, the multiple fractures that form a disposal domain (Bai *et al.,* 2010) would be created during the cuttings re-injection. The advantage of maintaining the injected cuttings within the multiple fractures of the disposal domain is to restrict the cuttings from environmentally protected regions. The accurate determination of these disposal regions is a critical role for the cuttings re-injection modeling.

11.6   FRACTURE PACKING IN UNCONSOLIDATED FORMATION

11.6.1   *Theoretical basis in simulation*

In the current reservoir stimulation, hydraulic fracturing techniques are no longer restricted to the brittle and lower permeability formations. In reality, they can apply to any type of formation in which the original formation permeability is not sufficiently large to aid a smooth hydrocarbon flow from the reservoir to the producing well. In addition to the application of hydraulic fracturing in ductile tight shale gas formations, as discussed in Section 11.4, hydraulic fracturing has been successfully used to stimulate the unconsolidated sandstone rocks in the past decades with the aid of the fracturing packing (or frac-pack) technique.

The frac-pack technique focuses on enhancing the propped fracture conductivity (i.e., fracture width) with the fully packed proppant to minimize the detrimental effect of fracture closing after the hydraulic fracture is created. As opposed to the conventional hydraulic fracturing that focuses on the propped fracture area (or fracture length for the fracture with restricted height growth), the frac-pack technique limits the development of the fracture area for the purpose of inflating the fracture width with the aid of substantial leakoff and tip screenout.

The tip screenout (TSO) can be artificially generated by progressively pumping high concentration proppant at a relatively high injection rate. Because it is difficult to propagate the fracture in soft formations, the substantially increased fracture width would compensate for the significantly reduced fracture area in the frac-pack operation that may create a similar fracture volume as that from hydraulic fracturing in brittle formations.

In addition to being used as an effective method for stimulating soft rock formations, the frac-pack method can be used for sand control as a sand exclusion device similar to gravel packing. The applications presented in this section, however, focus on reservoir stimulation using the frac-pack method only.

11.6.2   *An equivalent frac-pack model*

The key steps for creating a successful frac-pack include the following:

- Avoid any near-well clogging that may lead to premature screenout
- Initiate TSO by using a higher injection rate and introducing a larger leakoff at the fracture boundary (i.e., tip area)
- Place the proppant evenly to achieve back packing with the controlled TSO

The equivalent frac-pack model is achieved by using the conventional hydraulic fracturing simulator with a proper pumping schedule and the following effective input parameters:

- Larger injection rate [i.e., 20 bbl/min (3.18 m³/min) or greater,]
- Greater proppant concentration (i.e., final stage proppant concentration of greater than 15 lb/gal (1797.45 kg/m³) but less than 20 lb/gal (2396.6 kg/m³) while proppant ramping from a small value to a large value must be gradual)
- Greater formation leakoff percentage (i.e., between 80 and 99%)
- Sufficient stress contrast or permeability contrast between the pay zone and bounding layers (A general rule of thumb is that it needs a minimum of a few hundreds psi between the payzone and bounding layers for the fracture containment by the stress contrast.)

To identify the onset of TSO from the pressure response, Nolte and Smith (1979) designed a log-log plot of pressure vs. time, and claimed that a slope of 45° or greater angle between pressure and time is required for the TSO to occur. Fig. 11.9 provides an example of this criterion.

The onset of TSO occurs in the early stage of pumping under the effect of greater injection rate and relatively lower proppant concentration, which pushes the proppant to the tip area. Because the small space in the fracture tip cannot accept too much proppant, the clogging that occurs in the tip area hinders the further fracture propagation. As a result of pressure buildup under the continuing injection and ceased fracture growth, the fracture begins to expand along the width dimension. This expansion consequently creates more space within which to accept the proppant and results in the packing of proppant backward from the tip to the well. Fig. 11.10 illustrates the TSO and the frac-pack. The distributions in Fig. 11.10 are not conventional proppant concentrations (unit lb/ft²), but proppant ratio (i.e., ratio of proppant volume to injected volume, unitless). For the proppant concentration, the proppant mass per unit area is examined. Considering that the fracture is not two-dimensional, the effect of the fracture width in the third dimension on the solids packing is also crucial. As a result, the screenout scenario can be best represented by the value of the proppant ratio, as demonstrated in Fig. 11.10, rather than by the conventional proppant concentration.

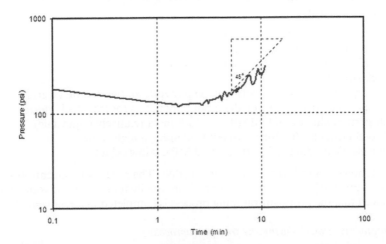

Figure 11.9. Net pressure vs. time in log-log scale to determine the onset of TSO (Nolte and Smith 1979; 1 psi = 0.0069 MPa).

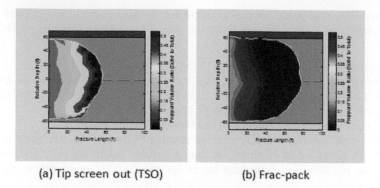

(a) Tip screen out (TSO)   (b) Frac-pack

Figure 11.10.   Tip screenout in early stage injection and frac-pack in late stage injection. The distributions are proppant ratio which is the ratio between injected proppant volume and injected total volume (1 ft = 0.3048 m).

Table 11.7.   An example input parameters for modeling frac-pack.

Parameter	Unit	Value	Unit (SI)	Multiply by
Reservoir pressure	psi	4,300	MPa	0.0069
Permeability	mD	100	m^2	9.87×10^{-16}
Porosity	%	28		
Fluid density	lb/ft^3	62.4	kg/m^3	16.0185
Fluid rheology constant n'		0.5		
Fluid rheology constant K'	lbf s^n/ft^2	0.056	kg s^n/m^2	4.8826
Leakoff coefficient	ft/min$^{1/2}$	0.02	m/min$^{1/2}$	0.3048
Reservoir thickness	ft	120	m	0.3048
Top of sand	ft	10,000	m	0.3048
Shale Young's modulus	psi	1,000	MPa	0.0069
Shale Poisson's ratio		0.29		

11.6.3   *Key input parameters for frac-pack modeling*

As mentioned previously, the critical factors to create the tip screenout and frac-pack procedure include the following:

- Relatively larger formation permeability
- Relatively greater injection rate
- Incrementally increasing proppant concentration
- Lower rock strength
- Relatively larger stress contrast between the pay zone and bounding layers
  Table 11.7 provides an example of the input parameters for frac-pack modeling.
- Reservoir permeability of 100 mD (9.87×10^{-16} m^2, a relatively high value)
- Leakoff coefficient of 0.02 ft/min$^{1/2}$ (0.0061 m/ min$^{1/2}$, a high value)
- Formation elastic modulus of 1,000 psi (6.9 MPa, a low value)

Table 11.8 shows a sample stage pumping schedule. The proppant concentration ramps up from 0 to 17.3 ppg (2073 kg/m^3) in 13 stages. From the incremental fluid volume injected, it can be deduced that the proppant ramping process is completed in a relatively short period.

11.6.4   *Fracture re-growth during the frac-pack process*

In numerical modeling, the TSO is interpreted as a stopping point for the fracture growth; in the simulation treatment, the fracture length and height reach a standstill when the TSO occurs. After that, the fracture width will expand under the continuing pressure buildup until the limiting injection pressure is reached, then the proppant is injected to fill the propped fracture. The pump will be shut down when the fracture is fully packed.

This TSO and frac-pack procedure in modeling can be either a misconception about what has happened in a real frac-pack procedure, or it can reflect a limitation by a numerical simulator to simulate the actual fracture response in a real frac-pack operation.

As a matter of fact, the fracture may re-grow in all three dimensions (i.e., along length, height, and width) under the continuous injection even after the TSO is reached. This process can be described as follows:

- Begin injection → TSO → fracture re-growth → further injection
- Start another loop until either the final injection volume is reached or the pumping pressure is too high to continue the injection
- End injection

As an example, Fig. 11.11, Fig. 11.12, Fig. 11.13, and Fig. 11.14 demonstrate the time-dependent responses of net pressure, fracture width, fracture height, and fracture length subjected to the continuous injection in the form of multiple TSOs and fracture re-growths.

The fracture re-growth during the frac-pack operation may occur in any direction (width, height, or length). These re-growths can occur to the fracture geometry individually, or simultaneously. Fig. 11.15 summarizes the scenarios shown in Fig. 11.11, Fig. 11.12, Fig. 11.13,

Table 11.8.   An example stage pumping schedule for modeling frac-pack.

Stage	Fluid volume (bbl)	Fluid volume (m³)	Proppant slurry concentration (ppa)	Proppant slurry concentration (kg/m³)
1	40.0	6.36	0	0
2	16.7	2.66	0.5	60
3	15.9	2.53	1.5	180
4	15.4	2.45	2.8	336
5	14.9	2.37	4.2	503
6	14.5	2.31	6.1	731
7	14.1	2.24	7.9	947
8	13.8	2.19	9.3	1114
9	13.5	2.15	12.4	1486
10	13.2	2.10	13.8	1654
11	13.0	2.07	15.2	1821
12	12.5	1.99	17.3	2073
13	100.0	15.90	17.3	2073

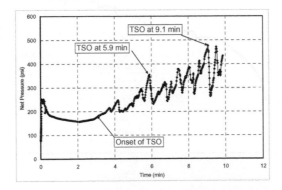

Figure 11.11.   Net pressure vs. injection time under continuous injection. It shows the pressure responses at onset and subsequent two TSOs during the injection. The TSO is determined by the Nolte-Smith method shown in Fig. 11.9 (1 psi = 0.0069 MPa).

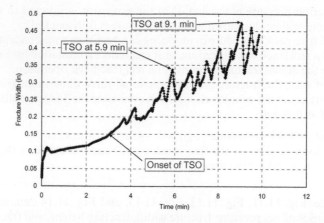

Figure 11.12.   Fracture width vs. injection time under continuous injection. It shows the fracture width re-growths at onset and subsequent two TSOs during the injection (1 in = 2.54 cm).

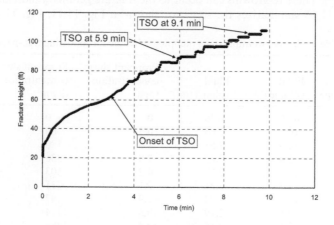

Figure 11.13.   Fracture height vs. injection time under continuous injection. It shows the fracture width re-growths at onset and subsequent two TSOs during the injection (1 ft = 0.3048 m).

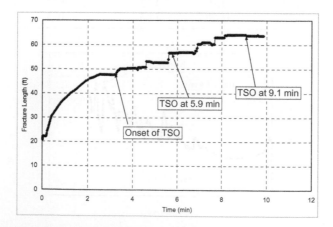

Figure 11.14.   Fracture length vs. injection time under continuous injection. It shows the fracture width re-growths at onset and subsequent two TSOs during the injection (1 ft = 0.3048 m).

and Fig. 11.14 in one plot, which indicates that the fracture re-growth occurs unevenly (i.e., at different times) to the specific direction (i.e., either width, height, or length).

Other observations from Fig. 11.15 are that the response of fracture width is similar to that of net pressure, whereas the responses of fracture height and length are insensitive to the response of net pressure. Therefore, it is true that the fracture width growth may dominate the frac-pack process. However, it is also true that fracture growths in height and in length are not negligible during the frac-pack process. The numerical capability to simulate the fracture re-growth in the frac-pack operation is consequently critical to a successful replication of the frac-pack process.

11.6.5 *Concluding remarks*

Like the simulation of cuttings re-injection modeling, when simulating the frac-pack procedure by means of hydraulic fracturing, the existing simulators for the conventional fracturing simulation can be used with some modifications to the pumping schedule and to some key input parameters. The most important modification of a conventional hydraulic fracturing simulator is to increase the proppant concentration incrementally to a high value

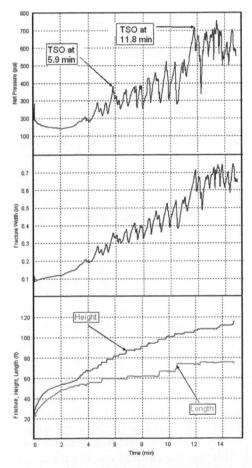

Figure 11.15. Comparison of net pressure and fracture geometric evolutions for a frac-pack operation. Fracture re-growths from the directions of width, height, and length may not occur at the same time (1 psi = 0.0069 MPa; 1 in = 2.54 cm; 1 ft = 0.3048 m).

to induce the TSO. Other important input parameters include relatively higher injection rate, greater formation permeability and leakoff capability, lower rock strength, and greater stress contrast between the pay zone and bounding layers (i.e., greater fracture height containment). Although the fracture growth dominates in the direction along the fracture width after the TSO, the fracture re-growth in all directions (i.e., width, height, and length) is a reality. It is not truthfully simulated, however, because of the numerical inability of some hydraulic fracturing simulators, especially from those that are not capable of simulating a 3D planar shape hydraulic fracture.

11.7   PRODUCED WATER RE-INJECTION

11.7.1   *Theoretical basis in simulation*

Produced water can come from several sources, including the following:

- Formation water.
- Injected water for waterflooding in the enhanced oil recovery.
- Condensed water in enhanced gas production.

The percentage of water production in hydrocarbon production can be quite high worldwide (approximately 75% [Farajzadeh 2004]). The remaining trace of oil (OIW) and generated solids (TSS) during the water-oil separation process may remain within the produced water. The composition of produced water is highly field-dependent or case-dependent. The produced water may be treated to remove some of OIW and TSS before being disposed. Produced water disposal by re-injection is one of the many disposal methods and is closely regulated and controlled by the local legislation agencies. Sometimes, the produced water can be reused for reservoir pressure maintenance.

Even with some treatment, the produced water may still be impure (e.g., containing residual hydrocarbons, heavy metals, radionuclides, inorganic species, suspended solids, and chemicals) that can result in the injectivity decline (near well) and formation damage (anywhere from well). The injectivity decline is related to the complicated near-wellbore effect, and can be the end result of formation damage from the produced water re-injection (PWRI). The case study presented here focuses only on examining the permeability reduction that leads to the particle plugging of the formation and plugging of the hydraulic fracture that eventually results in the injectivity decline.

In the modeling, two types of approaches are available:

- Matrix injection in which the injection pressure is not great enough to create the hydraulic fracture. The produced water enters the formation in the distributed fashion, and the flow within the porous media is primarily interstitial.
- Fracturing injection in which the injection pressure is great enough to exceed the in-situ stress, and a hydraulic fracture is generated. The produced water enters the formation near the well primarily through the hydraulic fracture in a channeled way.

11.7.2   *An equivalent produced water re-injection model*

It is difficult to determine the transport mechanism of produced water, regardless of whether it is through the matrix injection or through the fracturing injection; in some cases, the two mechanisms could be revealed by means of interpreting the injection pressure measurement data. This may be the primary reason that modeling techniques for PWRI have been left far behind that for CRI. Although the cuttings are injected into the formation as a slurry, it is easy to imagine that solids in the slurry would not travel far from the injection well. In contrast, the produced water with a small percentage of solids (e.g., TSS) may travel a long distance from the well.

Because the fracturing injection is no longer critical for disposing the produced water, it is assumed that the matrix injection is the primary process for the produced water re-injection modeling. In the equivalent **PWRI** model, several important flow and transport mechanisms influence the waste transport process:

- Cross-flow between the injection layer and neighboring layers.
- Injectivity decreases over the injection period as a result of the near-wellbore particle clogging.
- Solids deposition near the well as a retardation factor in the transport process.

11.7.3 *Numerical modeling of cross flow in produced water transport*

Based on the coupled poroelastic theory, finite element models were used to investigate the injection cases with the simplified geometries. The 2D plane strain geometry was attempted to visualize the cross-flow phenomena in the layered media where the point injections may imitate a line source along a specific vertical plane. Fig. 11.16 shows the mesh layout of a four-layer formation domain; Table 11.9 shows different permeability contrasts. In particular, Layers 1 (second layer) and 2 (third layer) are injection layers; Layers 3 (bottom layer) and 4 (top layer) are injection zone bounding layers.

The injection pressure is constant at 20 MPa along the left boundary of Layer 2 (more permeable layer), and is 10 MPa at Layer 1 (less permeable layer). In all other boundary locations, the constant pressure of 5 MPa is imposed.

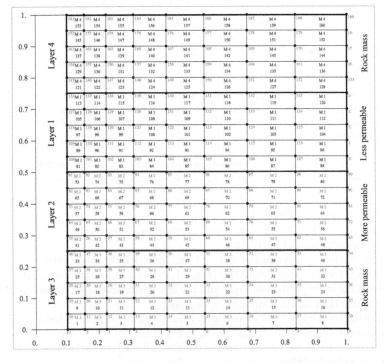

Figure 11.16.   Finite element mesh layout for a four-layer formation domain (cross-section view). Layers 1 and 2 are injection layers; Layers 3 and 4 are underburden and overburden, respectively. In each element, the information includes: a) node number (top left), b) material number (top middle) and c) element number (middle). The lateral axis is the distance from the injection location (meters) and the vertical axis is the relative distance (meters) from the lower bound of bottom layer (i.e., Layer 3).

Table 11.9.   Four-layer formation with different permeabilities.

Layer	Permeability coefficient	Unit	Note
1	0.001	m²/MPa s	M1: less permeable injection layer
2	0.1	m²/MPa s	M2: more permeable injection layer
3	0.00001	m²/MPa s	M3: bottom layer
4	0.00001	m²/MPa s	M4: top layer

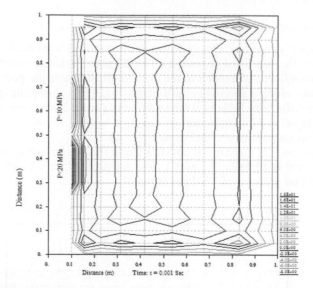

Figure 11.17.   Distribution of fluid pressure (MPa) in the four-layer porous media at 0.001 s after the injection. Injection occurs at middle layers. 20 MPa is the injection pressure at the more permeable layer (lower). 10 MPa is the injection pressure at the less permeable layer (upper).

The full historic development of cross-flow during the injection process is illustrated in Fig. 11.17 through Fig. 11.23. The following list includes a detailed description:

- The early time (i.e., t = 0.001 sec) pressure distribution is given in Fig. 11.17 in which the pressure is uniformly distributed.
- At later time (i.e., t = 0.1 sec), the pressure distribution shows the flow mainly occurring in the more permeable zone (Fig. 11.18).
- The flow channeling is sustained at time = 1 sec, in which the flow is still not significantly initiated within the less permeable layer (Fig. 11.19), except for some short-distance cross-flow between the more permeable layer and the less permeable layer, as well as between the more permeable layer and the adjacent rock mass.
- The substantial fluid leakoff (or cross-flow) from more permeable to less permeable layers may be apparent at time = 10 sec, as shown in Fig. 11.20.
- At time = 100 sec, the flow becomes more uniform and the large-area cross-flow is substantially reduced (Fig. 11.21).
- At later times (after t = 1000 sec in Fig. 11.22), the pressure distributions remain almost unchanged, signifying a quasi-steady state flow process in which only minimal cross-flow from the permeable layers to the rock masses to can be observed (time = 10,000 sec, in Fig. 11.23).

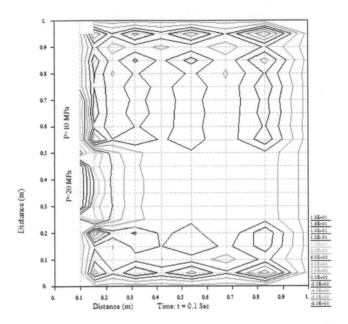

Figure 11.18.   Distribution of fluid pressure (MPa) in the four-layer porous media at 0.1 s after the injection. Injection occurs at middle layers. 20 MPa is the injection pressure at the more permeable layer (lower). 10 MPa is the injection pressure at the less permeable layer (upper).

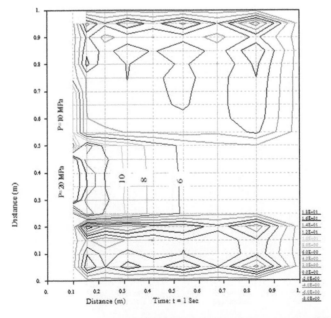

Figure 11.19.   Distribution of fluid pressure (MPa) in the four-layer porous media at 1 s after the injection. Injection occurs at middle layers. 20 MPa is the injection pressure at the more permeable layer (lower). 10 MPa is the injection pressure at the less permeable layer (upper).

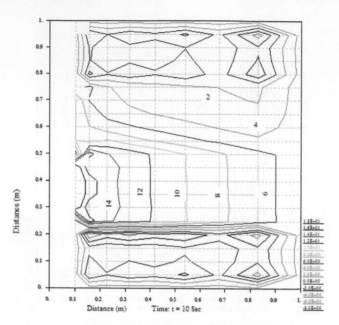

Figure 11.20. Distribution of fluid pressure (MPa) in the four-layer porous media at 10 s after the injection. Injection occurs at middle layers. 20 MPa is the injection pressure at the more permeable layer (lower). 10 MPa is the injection pressure at the less permeable layer (upper).

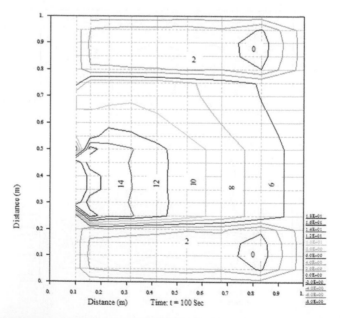

Figure 11.21. Distribution of fluid pressure (MPa) in the four-layer porous media at 100 s after the injection. Injection occurs at middle layers. 20 MPa is the injection pressure at the more permeable layer (lower). 10 MPa is the injection pressure at the less permeable layer (upper).

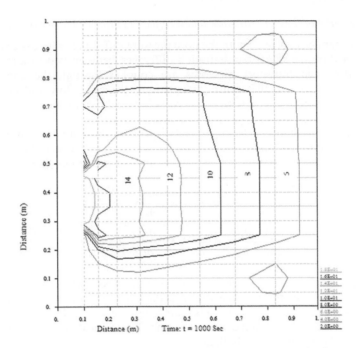

Figure 11.22.   Distribution of fluid pressure (MPa) in the four-layer porous media at 1,000 s after the injection. Injection occurs at middle layers. 20 MPa is the injection pressure at the more permeable layer (lower). 10 MPa is the injection pressure at the less permeable layer (upper).

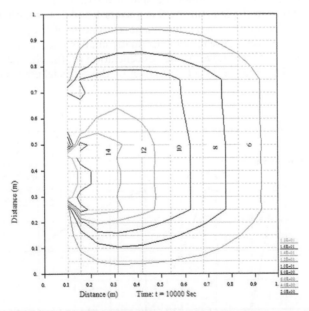

Figure 11.23.   Distribution of fluid pressure (MPa) in the four-layer porous media at 10,000 s after the injection. Injection occurs at middle layers. 20 MPa is the injection pressure at the more permeable layer (lower). 10 MPa is the injection pressure at the less permeable layer (upper).

11.7.4   *Analytical modeling of cross flow and its effect on produced water transport*

Cross-flow is generally associated with the pore pressure difference, whereas cross-transport is generally based on the solute concentration difference between two locations. However, because the flow and transport processes are coupled through the convection flow terms (i.e., flow rate, Q) in the transport equations, the cross-flow and cross-transport responses are intermingled. As a result, the cross-flow and cross-transport terms are used interchangeably.

For a two-layered medium, neglecting global dispersion, the coupled transport equations incorporating the solid retardation and cross-flow (or cross-transport), the governing coupled equations in radial plane transport geometry can be expressed as (Bai *et al.,* 2009):

$$-\frac{Q_1}{2\pi h_1 r}\frac{\partial c_1}{\partial r} = \phi_1\frac{\partial c_1}{\partial t} + K_{p1}\frac{\partial c_1}{\partial t} + \xi(c_1 - c_2) \tag{11.1}$$

$$-\frac{Q_2}{2\pi h_2 r}\frac{\partial c_2}{\partial r} = \phi_2\frac{\partial c_2}{\partial t} + K_{p2}\frac{\partial c_1}{\partial t} - \xi(c_1 - c_2) \tag{11.2}$$

where the subscripts 1 and 2 represent Layers 1 and 2, respectively; c is the solute concentration, Q is the flow injection rate, h is the layer thickness, r is the radial distance from the well center, Φ is the porosity, K_p is the solids retardation factor, ξ is the inter-layer transport coefficient., and t is the time from the initial injection. The scenario presented in Eq. 11.1 and Eq. 11.2 enables the injection from both layers.

The solids retardation factor K_p may be considered as a lumped retarding influence on the solute transport process. For simplicity, this influence is attributed to the deposition of solids in the fluid mixture into pore spaces, physically retarding the transport rate.

With a higher injection rate, smaller retardation factor, and higher porosity in Layer 1 than those in Layer 2, Fig. 11.24 shows the spatial distribution of solute concentrations (e.g., produced water) from the injection well in Layers 1 and 2. Cross-flow occurs between Layers 1 and 2. Layer 2 is prone to more significant retardation of the particle transport. The transport behavior for Layer 1 reveals some dramatic change occurring mainly in the later years. Between years 5 and 10 in Layer 1, the concentration profile changes from the S shape to the convex shape. This change reflects the effect of the artificial concentration boundary

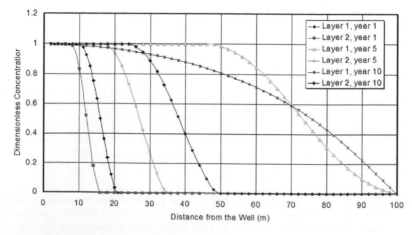

Figure 11.24.   Distribution of solute concentration in two layers with different injection rates and solids retardation factors; the solute travels farther in Layer 1 than in Layer 2.

at 100 m. Because of the higher injection rate and smaller retardation, solute (or produced water) transport is much farther in Layer 1 than in Layer 2.

11.7.5 *Concluding remarks*

Unlike the simulations of cuttings re-injection and frac-pack, the simulation of the produced water re-injection is more realistically performed by means of matrix injection; the fracturing injection may not be needed because the produced water contains low solid content, and its flow and transport process can be primarily conducted through interstitial porous spaces. In this circumstance, the important issues in the produced water re-injection modeling are associated with the ability to simulate the cross-flow between contrasting formation layers. The cross-flow between formation layers may be dictated by the following factors: injection rate, layer permeability and porosity, solid retardation factor, solute (i.e., produced water) concentration gradient, and layer geometries. The quality of the matrix injection can be severely affected by the wellbore injectivity, which tends to be influenced by the permeability degradation as a result of particle clogging in the near well region. As far as the modeling is concerned, the numerical simulation of PWRI is being challenged because of its complexity in modeling the process of combined flow and transport from matrix injection. The equivalent model for PWRI may have to be established over a hybrid numerical simulation of cross-flow and analytical modeling of solute (i.e., produced water) transport while the effect factors such as the injectivity decline as a result of near-well particle clogging along with the solids deposition as a retarding process should be additionally considered.

ACKNOWLEDGEMENTS

The author would like to thank Dr. John McLennan (University of Utah) and Sidney Green (TerraTek/University of Utah) for their enduring support and inspiration. The support from TerraTek during the course of the study is gratefully appreciated.

NOMENCLATURE

bbl	Barrel
bpm	Barrels per minute
c	Solute concentration
CRI	Cuttings re-injection
CRI-HF	Cuttings reinjection hydraulic fracturing
H	Layer thickness
HF	Hydraulic fracturing
IC	Injection center
Kp	Solids retardation factor
OIW	Oil in water
PWRI	Produced water re-injection
Q	Injection rate
R	Radial distance from well
t	Time from initial injection
TSG	Tight-shale gas
TSO	Tip screenout
TSS	Total suspended solids
Φ	Porosity
ξ	Inter-layer transport coefficient

REFERENCES

Abou-Sayed, A., Zaki, K., Wang, G., Meng, F. and Sarfare, M.: Fracture propagation and formation disturbance during injection and frac-pack operations in soft compacting rocks. Paper SPE 90656 presented at the SPE Annual Technical Conference and Exhibition, Houston, TX, USA, 26–29 September, 2004.

Anthony, B.: Hydraulic fracturing: a proven well completion technology for shale gas, Oklahoma Corporation Commission, NARUC meeting at Sacramento, CA, USA, 2010.

Bachman, R.C., Harding, T.G. and Settari, A.: Coupled simulation of reservoir flow, geomechanics, and formation plugging with application to high-rate produced water reinjection. Paper SPE 79695 presented at the SPE Reservoir Simulation Symposium, Houston, TX, USA, 3–5 February, 2003.

Bai, M., Green, S. and Suarez-Rivera, R.: Effect of leakoff variation on fracturing efficiency for tight shale gas reservoirs, *Proc. 40th U.S. Rock Mechanics Symposium*, Anchorage, AK, USA, 2005.

Bai, M., Morales, R.H., Suarez-Rivera, R. and Green, S.: Modelling fracture tip screenout and application for fracture height growth control. Paper SPE 83106 presented at the SPE Annual Technical Conference and Exhibition, Denver, CO, USA, 5–8 October, 2003.

Bai, M., McLennan, J.D., Buller, D. and Hagan, J.: Injection Modelling, Chapter 7, *SPE Solids Injection, Monograph 24*, Nagel and McLennan (eds), SPE (ISBN 978-1-55563-256-4), 2010.

Bai, M., McLennan, J., Guo, Q. and Green, S.: Cyclic injection modelling of cuttings re-injection, *Proc. 41th U.S. Rock Mechanics Symposium*, Golden, Colorado, USA, 2006.

Bai, M., McLennan, J.D. and Standifird, W.: Modelling fluid mixture transport and cross-flow in layered media, *Proc. 43th U.S. Rock Mech. Symp. and 4th U.S.-Canada Rock Mech. Symp.*, Asheville, NC, USA, 2009.

Clifton, R.J. and Abou-Sayed, A.S.: A variational approach to the prediction of the three-dimensional geometry of hydraulic fractures. Paper SPE 9879 presented at the SPE Low Permeability Symposium, Denver, CO, USA, 27–29 May, 1981.

Economides, M.J. and Nolte, K.G.: *Reservoir stimulation*. 3rd Edition. Wiley, 2000.

Farajzadeh, R: Produced water re-injection (PWRI) – an experimental investigation into internal filtration and external cake build up, PhD Thesis, Delft University, The Netherlands, 2004.

Gu, H. and Weng, X.: Critical for fractures crossing frictional interfaces at non-orthogonal angles. Paper ARMA-10-198 presented at the 44th US Symposium on Rock Mechanics, Salt Lake City, Utah, USA, 27–30 June, 2010.

Guo, Q and Geehan, T.: An overview of drill cuttings re-injection – lessons learned and recommendations, 11th Int. Petro. Environ. Conf., Albuquerque, NM, USA, 2004.

Haimson, B.C. and Fairhurst, C.: Initiation and extension of hydraulic fractures in rocks. *SPEJ*, 7:3 (1967), pp. 310–318.

Halliburton Consulting: Shale Development Guidebook, 2010.

Hubbert, M.K. and Willis, D.G.: Mechanics of hydraulic fracturing. *Trans. SPE-AIME* 210 (1957), pp. 153–168.

Khodaverdian, M. and McElfresh, P.: Hydraulic fracturing stimulation in poorly consolidated sand: mechanisms and consequences. Paper SPE 63233 presented at the SPE Annual Technical Conference and Exhibition, Dallas, TX, USA, 1–4 October, 2000.

Nolte, K.G. and Smith, M.B.: Interpretation of fracturing pressures. Paper SPE 8297 presented at the SPE Annual Technical Conference and Exhibition, Las Vegas, NV, USA, 23–26 September, 1979.

Sarout, J., Molez, L., Gueguen, Y. and Hoteit, N.: Shale dynamic properties and anisotropy under triaxial loading: experimental and theoretical investigations. *Physics and Chemistry of the Earth*, Elsevier, 1:7, 2006.

SI METRIC CONVERSION FACTORS

$$\text{bbl} \times 1.589\ 873 \quad \text{E} - 01 = \text{m}^3$$
$$\text{cp} \times 1.0 \quad \text{E} - 03 = \text{Pa} \cdot \text{s}$$
$$\text{ft} \times 3.048 \quad \text{E} - 01 = \text{m}$$
$$\text{gal} \times 3.785\ 412 \quad \text{E} - 03 = \text{m}^3$$
$$\text{in} \times 2.54 \quad \text{E} + 00 = \text{cm}$$
$$\text{Ibm} \times 4.535\ 924 \quad \text{E} - 01 = \text{kg}$$
$$\text{psi} \times 6.894\ 757 \quad \text{E} + 00 = \text{kPa}$$

Subject index

Multiphysics Modeling

Series Editors: Jochen Bundschuh & Mario César Suárez Arriaga

ISSN:1877-0274

Publisher: CRC/Balkema, Taylor & Francis Group

1. Numerical Modeling of Coupled Phenomena in Science and Engineering: Practical
   Use and Examples
   Editors: M.C. Suárez Arriaga, J. Bundschuh & F.J. Domínguez-Mota
   2009
   ISBN: 978-0-415-47628-7

2. Introduction to the Numerical Modeling of Groundwater and Geothermal Systems:
   Fundamentals of Mass, Energy and Solute Transport in Poroelastic Rocks
   J. Bundschuh & M.C. Suárez Arriaga
   2010
   ISBN: 978-0-415-40167-8

3. Drilling and Completion in Petroleum Engineering: Theory and Numerical Applications
   Editors: Xinpu Shen, Mao Bai & William Standifird
   2011
   ISBN: 978-0-415-66527-8

Printed and bound by CPI Group (UK) Ltd, Croydon, CR0 4YY

18/10/2024

01776254-0001